Techniques for Adaptive Control

Techniques for Adaptive Control

Edited by

Vance J. VanDoren, Ph.D., P.E.
**Consulting Editor, Control Engineering Magazine,
Oak Brook, Illinois**

An imprint of Elsevier Science

Amsterdam   Boston   London   New York   Oxford   Paris   San Diego
San Francisco   Singapore   Sydney   Tokyo

Butterworth–Heinemann is an imprint of Elsevier Science.

Copyright © 2003, Elsevier Science (USA). All rights reserved.

No part of this publication may be reproduced, stored in a retrieval system, or transmitted in any form or by any means, electronic, mechanical, photocopying, recording, or otherwise, without the prior written permission of the publisher.

♾ Recognizing the importance of preserving what has been written, Elsevier Science prints its books on acid-free paper whenever possible.

Library of Congress Cataloging-in-Publication Data

Techniques for adaptive control / edited by Vance J. VanDoren.
   p. cm.
Includes bibliographical references and index.
ISBN 0-7506-7495-4 (pbk. : alk. paper)
1. Adaptive control systems. I. VanDoren, Vance J.
TJ217 .T42 2002
629.8'36—dc21   2002074432

British Library Cataloguing-in-Publication Data
A catalogue record for this book is available from the British Library.

The publisher offers special discounts on bulk orders of this book.
For information, please contact:

Manager of Special Sales
Elsevier Science
200 Wheeler Road
Burlington, MA 01803
Tel: 781-313-4700
Fax: 781-313-4882

For information on all Butterworth–Heinemann publications available, contact our World Wide Web home page at: http://www.bh.com

10 9 8 7 6 5 4 3 2 1

Printed in the United States of America

CONTENTS

Contributors	viii
Preface	ix
Introduction *Vance J. VanDoren*	1
Chapter 1 **Adaptive Tuning Methods of the Foxboro I/A System** *Peter D. Hansen*	23
Controller Structure	26
Minimum Variance Control	31
Control by Minimizing Sensitivity to Process Uncertainty	33
Algebraic Controller Design for Load Rejection and Shaped Transient Response	35
Algebraic Tuning of a Controller with Deadtime	39
Robust Adaptation of Feedback Controller Gain Scheduling	44
Feedforward Control	47
Adaptation of Feedforward Load Compensators	48
Conclusion	52
Chapter 2 **The Exploitation of Adaptive Modeling in the Model Predictive Control Environment of Connoisseur** *David J. Sandoz*	55
Model Structures	58
Issues for Identification	62
Adaptive Modeling	69
Other Methods	73
Simulated Case Study on a Fluid Catalytic Cracking Unit	76
Conclusion	96

Contents

Chapter 3 Adaptive Predictive Regulatory Control with BrainWave — 99
Mihai Huzmezan, William A. Gough, and Guy A. Dumont

The Laguerre Modeling Method	101
Building the Adaptive Predictive Controller Based on a Laguerre State Space Model	105
A Laguerre-Based Controller for Integrating Systems	110
Practical Issues for Implementing Adaptive Predictive Controllers	114
Simulation Examples	120
Industrial Application Examples	125
Conclusion and Lessons Learned	142

Chapter 4 Model-Free Adaptive Control with CyboCon — 145
George S. Cheng

Model-Free Adaptive Control with CyboCon	145
Concept of MFA Control	145
Single-Loop MFA Control System	149
Multivariable MFA Control System	158
Anti-Delay MFA Control System	170
MFA Cascade Control System	173
Feedforward MFA Control	177
Nonlinear MFA Control	180
Robust MFA Control	183
MFA Control Methodology and Applications	186
MFA Control Methodology	186
The Inside of MFA	189
Case Studies	191

Chapter 5 Expert-Based Adaptive Control: ControlSoft's INTUNE Adaptive and Diagnostic Software — 203
Tien-Li Chia and Irving Lefkowitz

On-Demand Tuning and Adaptive Tuning	204
History and Milestone Literature	205
Adaptive Control Structure and Underlying Principles	207
Identification-Based Adaptive Control	209
Expert-Based Adaptive Control: ControlSoft's INTUNE	217
Concluding Observations	230

Chapter 6	**KnowledgeScape, an Object-Oriented Real-Time Adaptive Modeling and Optimization Expert Control System for the Process Industries** *Lynn B. Hales and Kenneth S. Gritton*	**233**
	Intelligent Software Objects and Their Use in KnowledgeScape	235
	Artificial Intelligence and Process Control	239
	Neural Networks	252
	Genetic Algorithms	256
	Documenting the Performance of Intelligent Systems	258
	Putting It All Together: Combining Intelligent Technologies for Process Control	261
	Results: Using Intelligent Control in the Mineral-Processing Industry	264
	Conclusion	266
	Appendix: Table of Artificial Intelligence Reference Texts	268
Author Index		271
Subject Index		273

CONTRIBUTORS

George S. Cheng, Ph.D.
President and Chief Technical Officer, CyboSoft, General Cybernation Group Inc., Rancho Cordova, California

Tien-Li Chia, Ph.D.
President, ControlSoft, Cleveland, Ohio

Guy A. Dumont, Ph.D.
Professor, Department of Electrical and Computer Engineering, University of British Columbia, Vancouver, Canada

William A. Gough, P.Eng.
Vice President, Universal Dynamics Technologies Inc., Richmond, British Columbia, Canada

Kenneth S. Gritton, Ph.D.
Director of Technology and Development, EIMCO Process Equipment Company, Salt Lake City, Utah

Lynn B. Hales, B.S.
General Manager, KnowledgeScape Systems, EIMCO Process Equipment Company, Salt Lake City, Utah

Peter D. Hansen, B.S., M.S., Sc.D.
Consultant, and Bristol Fellow (Retired), The Foxboro Co., Foxboro, Massachusetts

Mihai Huzmezan, Ph.D., P.Eng.
Assistant Professor, The University of British Columbia; President, MIH Consulting Group Ltd., Vancouver, Canada

Irving Lefkowitz, Ph.D.
Vice President, ControlSoft, Cleveland, Ohio

David J. Sandoz, Ph.D.
Director (Retired), Control Technology Centre Ltd., Manchester, United Kingdom

PREFACE

The nature of this book bears some explaining. First, the term "control" in the title refers to feedback controllers used in factories and processing plants to regulate flow rates, temperatures, pressures, and other continuous process variables. Discrete controls, computer numerical controls, financial controls, and TV remote controls are different subjects altogether.

Second, the genre of this book is not easily classified. It is not a "textbook" in the sense of lessons presented in a logical order for the edification of college students studying a particular subject. The Introduction does present an overview of adaptive control technology, but each of the subsequent chapters has been written by a different author as a stand-alone presentation of his favorite approach to the subject. The chapters are not otherwise intended to relate to each other and are presented in no particular order.

Nor is this a "handbook" where each chapter describes a different aspect of the same subject. Here the chapters all describe the same thing—adaptive control techniques available as commercial software products—but from radically different points of view. Some of the authors have even gone so far as to suggest that their techniques are not merely acceptable alternatives to the competition, but are superior for certain applications. Which of them is right is still a matter of debate.

This book could be considered a "survey" of adaptive control techniques, but not in the academic sense. The chapters have not been peer reviewed, they do not begin to cover the wide variety of adaptive controllers that have been developed strictly for academic purposes, nor have they been written according to academic standards. They are certainly technical and they contain plenty of equations to describe how each technique works, but simulations and real-life applications have been given more emphasis than detailed mathematical proofs.

Perhaps "trade show on paper" would be a more accurate description. Each author has contributed a chapter that describes his technology rather than a booth that displays it. However, the idea is the same—assemble related products side-by-side and let the user decide which would be best for his application. On the other hand, the techniques presented herein may or may not prove useful for any particular application. A disclaimer is therefore in order: *The authors of the following chapters are solely responsible for the accuracy and completeness of their respective claims. The editor of this book neither endorses nor guarantees any particular adaptive control product for any particular purpose.*

That said, I gratefully acknowledge the contributions made by each of my coauthors. Some have spent their entire professional careers reducing arcane control theories into usable products with practical applications. Without them, this book would be neither necessary nor possible.

Vance J. VanDoren

INTRODUCTION

Vance J. VanDoren

Broadly speaking, *process control* refers to mechanisms for automatically maintaining the conditions of a mechanical, chemical, or electrical *process* at specified levels and to counteract random *disturbances* caused by external forces. A process can be virtually any collection of objects or materials with measurable and modifiable characteristics such as, a car traveling at a certain speed, a batch of beer brewing at a certain temperature, or a power line transmitting electricity at a certain voltage. The conditions of a process are generally measured in terms of *continuous process variables*, such as flow rates, temperatures, and pressures that can change at any time.

In a basic process control system, a *sensor* measures the process variable, a computer-based *controller* decides how best to correct the *error* between the actual and desired measurements, and an *actuator* such as a valve or a motor carries out the controller's decision to force the process variable up or down. The resulting change is then remeasured by the sensor and the whole sequence of operations repeats in an ongoing *feedback* or *closed loop*.

TERMINOLOGY

The study of process control engineering can occupy an entire academic or professional career, and is therefore beyond the scope of this book. However, a vocabulary lesson is in order since many control engineers, including the authors of the following chapters, use different terms for the same concepts.

For example, the process variable is also known as the *controlled variable* since it is the object of the controller's efforts. But since that quantity is also a result of whatever physical phenomena are at work in the process, it is sometimes described as the *process output*. The signal that the controller sends to the actuators is sometimes called the *controller output* or the *process input* because the actuators in turn apply it to the process. Other authors refer to it as the *control effort*, the *corrective action*, or the *manipulated variable* since it represents the quantity that the controller can manipulate directly.

The desired value that the controller is attempting to achieve for a particular process variable is almost universally known as the *setpoint*, though it is occasionally called the *reference value*. The procedure that the controller employs to determine its next control effort is variously referred to as the *control law*, the *control algorithm*, or the *control equation*. In the same vein, an actuator that implements the controller's decision is sometimes called the *final control element*.

If the control law is an algebraic equation, it almost always includes several coefficients that can be set by the user to prescribe just how hard the controller is required to work at eliminating the error between the process variable and the setpoint. These *controller parameters* can be adjusted to match the controller's performance with the behavior of the process, much as a guitar string can be adjusted to produce just the right pitch when plucked by a human controller. This operation is therefore known as *tuning*, and the adjustable parameters are often called *tuning parameters* or *tuning constants*.

For example, the basic *proportional controller* uses a percentage of the last error as the next control effort, assuming that a larger error necessitates a larger control effort (and conversely). Exactly what percentage or *gain* the controller should use to multiply with the error to compute the control effort is a matter of tuning. A higher gain would be appropriate for a sluggish process, whereas a lower gain would be necessary to prevent over-correcting a process that is more sensitive to the controller's efforts.

CONTENTS

Further details of how process controllers work and how their control laws are selected and tuned are left to textbooks on the subject. It will be assumed hereafter that the reader is a practicing control engineer or technician with at least a basic understanding of process controllers and their use in industrial applications.

What will be presented in the following chapters are several approaches to a particular process control technique called *adaptive control*. Although every process controller is "adaptive" in the sense that it changes its output in response to a change in the error, a truly adaptive controller adapts not only its output, but its underlying control strategy as well. It can tune its own parameters or otherwise modify its own control law so as to accommodate fundamental changes in the behavior of the process.

An adaptive proportional controller, for example, might increase its own gain when it observes that the process has become slow to respond to its control efforts. This could help maintain tight control over a process that experiences variable sensitivity, such as a heat exchanger. As sediment deposits begin to inhibit the exchange of heat, the controller might compensate by becoming more aggressive with a larger gain. Conversely, if the controller ever observes that its efforts have become too aggressive for the process (as would be the case immediately after a cleaning of the heat exchanger), it would reduce its own gain again.

Hundreds of techniques for adaptive control have been developed for a wide variety of academic, military, and industrial applications. Arguably the first rudimentary adaptive control scheme was implemented in the late 1950s using a custom-built analog computer (Kalman, 1958). Many "self-tuning" and "auto-tuning" techniques have been developed since then to automate the tuning procedures that an engineer would otherwise need to complete manually when commissioning a loop. Indeed, most commercial controllers today include some form of self-tuner or auto-tuner as a standard feature. However, these are generally one-shot operations that are useful at startup, but not necessarily for continuously updating the performance of the controller.

COMMERCIAL ADAPTIVE CONTROLLERS

Very few adaptive controllers capable of updating their control strategies *online* (i.e., while the process is running) have ever been commercialized as off-the-shelf software products. Two examples are BciAutopilot from Bachelor Controls (www.bciauto pilot.com), designed specifically for cooking extrusion applications, and the more general purpose QuickStudy from Adaptive Resources (www.adaptiveresources. com). Six others are the subject of this book:

- Chapter 1: EXACT, from The Foxboro Company (www.foxboro.com)
- Chapter 2: Connoisseur, also from Foxboro
- Chapter 3: BrainWave, from Universal Dynamics Technologies (www. brainwave.com)
- Chapter 4: CyboCon, from CyboSoft (www.cybocon.com)
- Chapter 5: INTUNE, from ControlSoft (www.controlsoftinc.com)
- Chapter 6: KnowledgeScape from KnowledgeScape Systems (www.kscape. com)

Each chapter has been written by the product's developers with the goal of presenting the functionality of their respective controllers along with some evidence of why or whether they work as claimed. Overt commercialism has been kept to a minimum, though several of the contributors have understandably demonstrated a measure of pride in their work. Other adaptive control products have no doubt been overlooked in the preparation of this book, and their developers are invited to contact the editor for inclusion in future editions.

The specific contents of each chapter vary widely since some contributors have gone into greater detail than others. Some have presented not only descriptions of their products, but background information about industrial process controls in general and adaptive controllers in particular. Others have focused on just the elements of their products that are particularly interesting, new, or unique. These disparities can be attributed to issues of technical complexity, the limited space allocated to each chapter, and most especially to the confidentiality of the contributors' trade secrets. These are commercial products, after all.

On the other hand, virtually every chapter delves into the technical details of process control theory to one degree or another. Nontechnical readers may just have to take it on faith that the theoretical results claimed by the contributors are true. Readers more familiar with process control theory may wish to consult the references and additional reading presented at the end of each chapter for further details, though not all of the techniques presented herein have been widely published in academic circles.

Nontechnical readers will also find examples of how well these products have worked in simulations and in actual practice. After all, the real proof of whether any technology has any practical value is whether or not it actually works. The nontechnical reader is advised to skip the theoretical sections and review the following chapter sections.

- Chapter 1: "Robust Adaptation of Feedback Controller Gain Scheduling" and "Adaptation of Feedforward Load Compensators"
- Chapter 2: "Simulated Case Study on a Fluid Catalytic Cracking Unit"
- Chapter 3: "Simulation Examples" and "Industrial Application Examples"
- Chapter 4: "Case Studies"
- Chapter 6: "Results Using Intelligent Control in the Mineral-Processing Industry"

PROBLEMS WITH PID

Traditional nonadaptive controllers are generally "good enough" for most industrial process control applications. The ubiquitous proportional-integral-derivative controller or *PID loop* is especially cheap and easy to implement. And though its operations are somewhat simplistic by the standards of modern control theory, a

PID loop can be remarkably effective at keeping the process variable close to the setpoint.

The simplicity of the PID controller also makes it fairly easy to understand and easy to diagnose when it fails to perform as desired. Tuning a PID controller is a relatively straightforward operation that can be accomplished with a few empirical tests that have remained essentially unchanged since the 1940s (Ziegler and Nichols, 1942). There are also a variety of well-developed techniques for extending the effectiveness of PID loops in more challenging applications such as gain scheduling for setpoint-dependent processes and the Smith predictor for deadtime-dominant processes.

However, even with these enhancements a PID controller leaves considerable room for improvement. Once tuned, it can only control the process it started with. If the behavior of the process changes appreciably after startup, the controller may no longer be able to counteract the error when a *load* disturbs the process variable. If the mismatch between the process behavior and the controller's original tuning becomes particularly severe, the closed-loop system may even become *unstable* as the controller alternately overcorrects, then undercorrects the error ad infinitum.

The traditional fix for coping with time-varying process behavior is to start over and manually retune the loop whenever its performance degrades. That may not be particularly difficult, but repeatedly tuning and retuning a loop can be tedious and time consuming, especially if the process takes hours to respond to a tuning test. Tuning rules also require at least some training to apply properly, so many PID controllers end up poorly tuned when implemented by inexperienced users. In extreme cases, plant operators will deactivate a poorly tuned controller when a disturbance occurs, then reactivate it once they've dealt with the disturbance manually. That strategy defeats the very purpose of feedback control.

ADVANTAGES OF ADAPTIVE CONTROL

Convenience is one of the most compelling reasons to replace PID loops with adaptive controllers. A controller that can continuously adapt itself to the current behavior of the process relieves the need for manual tuning both at startup and thereafter.

In some cases, manual retuning may not even be possible if the behavior of the process changes too frequently, too rapidly, or too much. A setpoint-dependent or *nonlinear* process can be particularly difficult to control with a fixed-parameter controller since it reacts differently to the controller's efforts depending on the current value of the setpoint.

A pH process, for example, becomes more efficient near a neutral pH level, requiring less titrant to achieve a given change in the pH. It is possible to equip a traditional controller with a different set of tuning parameters for each possible value of the setpoint (a strategy known as *gain scheduling*), but each set has to be manually adjusted. An adaptive controller can perform that chore automatically.

Adaptive controllers can also outperform their fixed-parameter counterparts in terms of efficiency. They can often eliminate errors faster and with fewer fluctuations, allowing the process to be operated closer to its constraints where profitability is highest. This is particularly advantageous in industries such as petrochemicals and aerospace where every ounce of performance counts for reasons of profits or safety.

DISADVANTAGES OF ADAPTIVE CONTROL

On the other hand, adaptive controllers are much more complex than traditional PID loops. Considerable technical expertise is required to understand how they work and how to fix them when they fail. The average plant engineer and his or her colleagues in operations are probably not going to understand such arcane adaptive control concepts as:

- BrainWave's weighted network of orthonormal Laguerre functions
- CyboCon's multilayer perceptron artificial neural network
- Connoisseur's radial basis function models

Fortunately, they don't have to. Commercial adaptive controllers are generally designed to make the technical details of their operations transparent to the user. And though it is the purpose of this book to explain at least the basic theories behind each technique, it really isn't necessary to fully understand an adaptive controller in order to use it. The details presented herein are intended to help justify the authors' claims and to provide a basis for comparing the relative merits of each technique.

That's not to say that adaptive controllers have been universally accepted as reliable and trustworthy. The plant's engineers may be convinced that the adaptive controllers they've selected work as promised, but the operators may not be as willing to trust the developers' incomprehensible technology. As a result, adaptive controllers are often relegated to supervisory roles where they do not control the process directly but generate setpoints for other traditional controllers to implement. Sometimes the adaptive controllers are further limited by constraints on their setpoint selections, and sometimes they end up disabled altogether.

THE IDEAL ADAPTIVE CONTROLLER

So just how good would an adaptive controller have to be to be considered absolutely trustworthy? Ideally, it would be able to track every setpoint change and counteract every disturbance given little or no input from the operators other than an objective to meet. It would operate in "black box" mode, observing the process and learning everything necessary to control it. It would be sufficiently *robust* to be unaffected by changes in the behavior of the process, or at the very least capable of changing its control strategy to accommodate them. Finally, it would be able to meet all of these objectives and bring the process variable into line with the setpoint moments after startup.

Clearly, no controller could do all that. For starters, it couldn't possibly select its control actions intelligently until it had already learned (or been told) something about the process. There are several ways to educate a controller. BrainWave and CyboCon ask the operators for hints based on existing knowledge or assumptions about the process. For example:

- Which process variable is most responsive to each of the controller's outputs?
- Does the process variable increase when the controller output increases or does it decrease?
- What is the time scale of the process? Does it take seconds or hours for the process variable to change by 10%? Does the process respond to control actions applied moments ago or hours ago? Which of those intervals is longer?

Alternately, the controller could answer these questions itself by conducting its own empirical tests on the process before startup. EXACT and Connoisseur both offer this option. Or, the controller could try its best guess at an initial control effort and deduce some kind of pattern from the results. However it is accomplished, some kind of manual or automatic *identification* operation is an indispensable first step toward effective control (adaptive or otherwise).

Then there's the issue of how the controller should combine its observations with the information supplied by the operators to design its own control strategy. There are as many answers to that question as there are control laws and tuning techniques. In fact, academic research is still ongoing in the area of self-modifying controller designs as well as automatic process identification.

Nonetheless, each of this book's contributors has come up with an approach that he thinks approximates the ideal adaptive controller and has turned it into a commercial product. Each works differently, but all fall into one of three basic categories:

- *Model-based* adaptive control
- *Model-free* adaptive control
- *Expert system* (or *rule-based* or *artificially intelligent*) adaptive control

although these technologies do overlap.

For example, BrainWave relies on a process model to create a suitable control law, but it uses several rules to determine when it is likely to have sufficient data to create the model correctly. KnowledgeScape can be configured to make use of a process model as well, and INTUNE can construct one just for informational purposes, but both rely primarily on expert systems to make their control decisions (albeit in vastly different ways). EXACT uses a process model and analytical tuning as the basis for designing its own control law but occasionally resorts to an expert systems approach when all else fails. Connoisseur is also model-based, but it works best when it can interact with an actual expert.

Basic Concepts

Traditional fixed-parameter controllers are often designed according to model-based control theory, using the process's historical behavior to predict its future. The historical behavior is represented by a mathematical model that describes how the inputs to the process have affected its outputs in the past. Assuming the same relationship will continue to apply in the future, the controller can then use the model to select future control actions that will most effectively drive the process in the right direction. It generally does so using a control law based on the model parameters. This can be as simple as a PID loop with tuning parameters computed from the model parameters algebraically or as complex as a calculus-based *model-predictive control* scheme.

Adaptive model-based controllers such as EXACT, BrainWave, and Connoisseur take that concept one step further. They generate their models automatically from historical data. Not only is this convenient when compared to designing controllers by hand, it permits ongoing updates to the model so that in theory the controller can continue to predict the future of the process accurately even if its behavior changes over time.

The one and only "model-free" adaptive controller described in this book—Cybo-Con—also derives its control law from an analysis of historical data, but without first creating a model of the process. In that sense it is similar to a traditional PID controller whose control law includes the current error, the sum of past errors, and how fast the error is changing. However, CyboCon's control law extracts considerably more detailed information from the last N error measurements so as to adapt to the current behavior of the process.

An expert system controller mimics the actions of experienced operators and engineers by tweaking the process or adjusting the controller just as they would, using their own rules. However, most expert system controllers are not adaptive since their control rules are fixed by the experts who programmed them. It would theoretically be possible to modify or add to those rules online, but that would require the ongoing involvement of an expert, and that's not the point of adaptive control.

KnowledgeScape, on the other hand, can adapt without changing its rule set. It uses a predictive process model so that the rules can be applied to the future as well as the present conditions of the process. And since that model can be updated online with recent process data, KnowledgeScape can adapt to changes in the behavior of the process. INTUNE also makes do with a fixed set of expert rules by using them not to manipulate the controller's output directly, but to adjust the tuning parameters of a traditional controller.

MODEL-BASED TECHNIQUES

Certainly the most common of these three techniques, and arguably the most obvious approach to adaptive control, is based on mathematical models of the process. Four of the six techniques described herein use process models in one fashion or another.

Given a model of the process, it is relatively easy for a controller to design an effective control law just as a control engineer would do when designing a traditional controller by hand. After all, an accurate model that can correctly predict the future effects of current control efforts contains all the mathematical information that the controller needs to select its control actions now so as to produce the desired process outputs in the future.

There are hundreds of techniques already available for translating model parameters into control laws, depending on the specific performance objectives the controller is required to meet. The hard part of model-based adaptive control technology is generating or *identifying* the model. There are three basic approaches to model identification:

- First principles
- Pattern recognition
- Numerical curve fitting

First principles were once the basis on which all model-based controllers were designed. They typically consisted of first- or second-order differential equations relating the present process output to its previous inputs and the derivatives thereof. These were especially practical for small-scale applications where enough was known about the process to analyze its behavior according to the laws of chemistry, physics, thermodynamics, etc.

MODERN ALTERNATIVES

First principles models are still used extensively today, but some modern processes (especially in the petrochemical and food industries) are so large and complex that their governing principles are too convoluted to sort out analytically. It may be clear from the general behavior of the process that it is governed by a differential equation of some sort, but the specific parameters of the model may be difficult to derive from first principles.

Pattern recognition was one of the first alternatives proposed to handle this situation. By comparing patterns in the process data with similar patterns characteristic of known differential equations, the controller could deduce suitable parameters for the unknown process model. Such patterns might include the frequency at which the process output oscillates as the controller attempts to counteract a disturbance or the rate at which the process output decays when the setpoint is lowered.

Pattern recognition techniques have succeeded in reducing the model identification problem to a matter of mathematics rather than physical principles, but they have their limitations as well. There's no guarantee that the process will demonstrate the patterns that the controller is programmed to recognize. For example, the EXACT controller looks for decaying oscillations in the process output after a disturbance. It deduces the process model by analyzing the size and interval between successive peaks and troughs. But if that response is not oscillatory or if the oscillations do not decay, it has to resort to an alternative set of expert rules to compute the model parameters.

Another alternative to first principles modeling is to compute the parameters of a generic equation that best fits the process data in a strictly numerical sense. Such empirical models are convenient in that they require no particular technical expertise to develop and they can be updated online for the purposes of adaptive control.

CURVE-FITTING CHALLENGES

However, numerical curve-fitting may not be able to capture the behavior of the process as accurately as first principles, especially in the presence of measurement noise, frequent disturbances, or nonlinear behavior. There's also a more insidious risk in relying on an empirical model for adaptive control: It can fit the data perfectly, yet still be wrong.

This problem is easy to spot when the input/output data is all zeros while the process is inactive. Any equation would fit that data equally well, so the modeling operation can simply be suspended until more interesting or *persistently exciting* data becomes available. The real trouble starts when the process becomes active, but not quite active enough. Under those conditions, the mathematical problem that must be solved to determine the model parameters can have multiple solutions. Worse still, it's generally not obvious whether the controller has picked the right solution or not.

Fortunately, there are ways to work around the persistent excitation problem and the spurious results it can cause. Some adaptive controllers will simply generate their own artificial disturbances (typically a temporary setpoint change) in order to probe the process for useful input/output data. Others will wait for naturally occurring disturbances to come along. BrainWave and the EXACT controller can do both. BrainWave can also be configured to add a *pseudo random binary sequence* (PRBS—an approximation of white noise) to the existing setpoint. This approach attempts to elicit useful data from the process without disturbing normal operations "too much." Connoisseur also has a PRBS function as well as the option to apply a setpoint change to the process or wait for a naturally occurring disturbance to stimulate the process.

Which of these is the "best" approach depends largely on the application. A particularly critical process that cannot be disturbed without jeopardizing profits, safety, or downstream operations would be a poor candidate for an adaptive controller that applies artificial disturbances. In such cases, the operators would typically prefer to take time to tune their loops by hand (and then only when absolutely necessary) rather than allow a controller to periodically ruin a production run just so it could learn what they already know. On the other hand, batch-oriented processes that switch from one setpoint to another as a matter of course would be easy to monitor for their responses to setpoint changes.

MORE CHALLENGES

A related problem can occur when the input/output data is flat because the controller has been successful at matching the process output to the setpoint. Should something happen to alter the process during that period of inactivity, the process's subsequent behavior may well differ from what the controller expects. Unless the controller first manages to collect new process data somehow, it could be caught completely off guard when the next natural disturbance or setpoint change comes along. It would most likely have to spend time identifying a new model before it would be able to retake control of the process. In the interim, errors would continue to accumulate and the performance of the control system would degrade. It could even become unstable, much like a PID controller with tuning that no longer matches the process.

On the other hand, this tends to be a self-limiting problem. Any fluctuations in the input/output data that result from this period of poor control would be rich with data for the modeling operation. The resulting model may even be more accurate than the one it replaces, leading ultimately to better control than before.

Modeling a process while the controller is operating poses yet another subtle problem. The mathematical relationship between the process's input and output data will be governed not only by the behavior of the process, but by the behavior of the controller as well. That's because the controller feeds the process output measurements back into the process as inputs (after subtracting the setpoint), giving the control law a chance to impose a mathematical relationship on the input/output data that has nothing to do with the behavior of the process itself.

As a result, an adaptive controller will get inaccurate results if it tries to identify the process model from just the raw input/output data. It has to take into account the input/output relationship imposed by the controller as well as the relationship imposed by the process. Otherwise, the resulting process model could turn out to be the negative inverse of the controller. Connoisseur gets around this problem by filtering its input/output data so as to distinguish the effects of the controller from the effects of the process.

Then there's the problem of noise and disturbances imposing fictitious patterns on the process outputs. A load on the process can cause a sudden change in the output measurements, even if the process inputs haven't changed. Sensor noise can cause an apparent change in the process variable simply by corrupting the sensors' measurements. Either way, an adaptive controller collecting input/output data at the time of the noise or the disturbance will get an inaccurate picture of how the process is behaving.

Most adaptive controllers work around the disturbance problem the way BrainWave does, by collecting data only while the process is in a steady state—that is, after it has finished responding to the last disturbance or setpoint change. The effects of measurement noise can also be mitigated by applying statistical filters to the raw measurements á la Connoisseur or by employing a modeling procedure that is unaffected by noise.

STILL MORE CHALLENGES

Those are some of the more challenging obstacles that model-based adaptive controllers face. Here are three more:

- *Input and output constraints.* Without a first-principles model of the process at hand, it may be difficult to determine where the process inputs and outputs will go on their way to the desired steady state. That in turn makes it more difficult for an adaptive controller to choose its control efforts so as to avoid the *constraints* of the process; that is, the maximum and minimum values allowed for the control efforts and the process variable. Constraints are often imposed on a process to keep it operating within a safety zone and to prevent the actuators from working themselves to death.
- *Tuning the tuner.* Although adaptive controllers are designed to tune or otherwise adapt themselves, they still need some guidance from the operators on how to do so. For example, BrainWave requires parameters to specify how hard and how fast its model identifier is to work and how many model components to use. The "correct" values for these and other parameters required for optimal adaptation are generally not obvious. Fortunately, the developers can usually provide rule-of-thumb values for the controller's operational parameters. They may not be optimal, but they'll work.

- *Keeping current.* If the model identifier is to track ongoing changes in the process, it has to discount older data in favor of more current observations. How it "forgets" ancient history (or more precisely, the time frame over which forgetting occurs) can significantly change the results in nonintuitive ways. On the other hand, some controllers such as Connoisseur will not forget an old model just because the process has been inactive for a while. Instead, they will reidentify the model when the next disturbance occurs.

POPULAR NONETHELESS

In spite of these and a host of related challenges, empirical models based on numerical curve-fitting have become a mainstay of many (perhaps even most) adaptive controllers in both academic and industrial applications. Such widespread interest in this field has led to the development of a dizzying array of model types and curve-fitting techniques. For example:

- Autoregressive moving average (ARMA) difference equations
- Radial basis functions
- Laguerre orthonormal functions
- Transfer functions
- Differential equations
- State equations
- Step/impulse response models
- Neural networks

These are some of the models employed just by the subjects of this book. Numerical curve-fitting techniques are more numerous still. Even the venerable least squares algorithm has multiple incarnations including "recursive," "iterative," and "partial." Just how and why these work is the subject of graduate study in mathematics and process controls, though several examples are explained in Chapter 2, in the sections entitled "Issues for Identification," "Adaptive Modeling," and "Other Methods," and in Chapter 5, in the section entitled "Model Identification (Identifier Block)."

Suffice it to say that each model type and each curve-fitting technique has its advantages and disadvantages, most of which are more mathematical than intuitive. Each also presents a myriad of options for selecting its mode of operation and making assumptions about the process. There is no "one size fits all" technique.

A MODEL-FREE TECHNIQUE

Nor are model-based techniques universally considered to be the best approach to adaptive control. After all, creating a model does not actually add any new information to the input/output data that every controller collects anyway. It certainly organizes the raw data into a convenient form from which a control law can be derived, but it should theoretically be possible to translate the input/output data directly into control actions without first creating any process model at all.

In fact, one of the earliest attempts at an adaptive controller using digital computer technology did just that (Åström and Wittenmark, 1973). It computed the control law directly from the input/output data. A model was implicit in the formulation of the control law, but it was never explicitly identified.

Likewise, CyboCon skips the modeling step and all of the problems that go with it. Instead of creating an input/output model of the process, CyboCon looks for patterns in the recent errors. This *learning* algorithm produces a set of gains or *weighting factors* that are then used as the parameters for the control law. It increases the weighting factors that have proven most effective at minimizing the error while decreasing the others. The weighting factors are updated at each sampling interval to include the effects of the last control action and recent changes in the process behavior.

It could be argued that the weighting factors implicitly constitute just another form of process model. Perhaps, but the weighting factors do not converge to values with any particular physical significance. They change when the behavior of the process changes, but their individual values mean nothing otherwise. Furthermore, the weighting factors in the control law can legitimately converge to values of zero. In fact, they do so every time the process becomes inactive. That in turn produces a zero control effort, which is exactly what is needed when the error is already zero; that is, when there are no disturbances to counteract nor any setpoint changes to implement.

PROS AND CONS

Arguably the most significant advantage of this strategy is that it avoids the trade-off between good modeling and good control that plagues most model-based techniques. When the process is inactive, CyboCon doesn't continue to look for meaning among the flat-line data. It simply attempts no corrective actions and continues waiting for something interesting to happen.

Academics will also appreciate CyboCon's *closed-loop stability conditions*, which turn out to be fairly easy to meet. Under these conditions, CyboCon will always be able to reduce the error without causing the closed-loop system to become unstable. That's a hard promise for an adaptive controller to make. For most model-based techniques it is possible to specify conditions under which an accurate model will eventually be found and how the closed-loop system will behave once the model is in hand. However, it is not generally possible to determine exactly how the closed-loop system will behave in the interim while the model is still developing (though BrainWave is a notable exception). The developers of Connoisseur recognize this fact and strongly recommend that modeling be conducted offline if at all possible or for short periods online under close operator supervision.

On the other hand, CyboCon isn't exactly the Holy Grail of adaptive control, either. Perhaps its biggest drawback is its virtually unintelligible control strategy. Even CyboCon's developers can't explain exactly what it's doing minute by minute as it generates each successive control effort. Only the end results are predictable. Furthermore, CyboCon's technology departs so dramatically from classical and even modern control theory that there are just a handful of academics and even fewer practicing engineers who actually understand why and how it works. Most users will simply have to assume that it does.

RULE-BASED TECHNIQUES

Although model-based and model-free techniques differ in their use of process models, they are similar in the sense that both use mathematical relationships to compute their control actions. Rule-based controllers, on the other hand, use qualitative rather than quantitative data to capture past experience and process history. That information combined with knowledge of the current state of the process is what allows the controller to choose a proper course of action.

There are essentially two ways to use expert rules for adaptive control, both of which are more heuristic than analytical. An "expert operator" controller such as KnowledgeScape manipulates the actuators directly. It acts like an experienced operator who knows just which valves to open and by how much. The rules rather than a mathematical equation serve as the control law.

An "expert engineer" controller such as INTUNE uses a traditional control equation, but tunes its parameters according to a set of expert rules. This could be as simple as applying the closed-loop Ziegler–Nichols tuning rules to a PID controller or as complicated as a home-grown tuning regimen developed over many years of trial and error with a specific process. The rules incorporate the expert engineer's tuning abilities rather than the expert operator's skill at manually controlling the process.

The format for such rules can vary widely, though they usually take the form of logical cause-and-effect relationships such as IF–THEN–ELSE statements. For example, the expert operator rules for a cooling process might include "IF the process temperature is above 100 degrees THEN open the cooling water valve by an additional 20%." An expert engineer rule might be "IF the closed-loop system is oscillating continuously THEN reduce the controller gain by 50%."

FUZZY LOGIC

The foregoing are examples of *crisp* rules that rely on conditions that are either entirely true or entirely false. *Fuzzy* rules deal with conditions that can be partially true and partially false. *Fuzzy logic* provides a computational mechanism for evaluating the results of a fuzzy rule and combining multiple rules to create complex logical relationships.

Fuzzy rules for the cooling process might include "IF the process temperature is moderately high THEN open the cooling water valve a little more." The terms "moderately high" and "a little more" would be defined as relatively true on a scale of 0 to 1, depending on just how high the temperature is currently and how far the valve has already been opened.

KnowledgeScape uses crisp as well as fuzzy rules to decide what control actions to take next. It also uses a neural network to model the process, allowing it to apply its expert rules not only to the current process conditions but to the future as well. It could decide "IF the process is *going* to be too hot, THEN start the cooling process *now*."

Note that the rules for a KnowledgeScape controller are user-defined specifically for each process. Its neural network model, on the other hand, is sufficiently generic to be automatically adapted to a broad class of processes. With INTUNE, the tuning rules themselves are the generic elements that allow the controller to adapt itself to the current behavior of the process.

MORE PROS AND CONS

Rule-based controllers can offer several advantages over traditional control techniques. Given a sufficiently encompassing set of rules, expert systems in general may be able to "reason" and perhaps even draw nonobvious conclusions from an incomplete and sometimes inaccurate sets of facts. In process control applications, this could result in combinations of control actions that no one ever thought of before. A rule-based controller may even uncover a better solution that had gone unnoticed simply because the process has always been run in a certain way. Expert systems generally do not respect tradition.

Rule-based controllers are also particularly easy to expand and enhance. Individual rules can be added or modified without revising the rest of the current set. This generally can't be done automatically, but it does make a rule-based controller especially flexible. Furthermore, if every new rule makes sense by itself and does not directly contradict any of the existing rules, the overall control strategy can be much easier to validate than an equally complex equation-based control strategy.

Expanding a model-based controller is generally not as easy since changing to a new model format generally requires starting again from scratch (though once again BrainWave is a notable exception). Rule-based controllers also have the advantage of being unaffected by the persistent excitation problem since most of them don't use process models to begin with. In fact, INTUNE's developers evolved from a model-based to a rule-based adaptive control strategy in large part to avoid the problems inherent with online process modeling, but also to achieve more efficient and reliable tuning in general.

On the other hand, the inexact nature of rule-based control is a double-edged sword. It frees the controller from some of the mathematical limitations suffered by model-based techniques, but it also makes stability and convergence difficult to assess. There are no mature mathematical principles available to determine when, how, or even if the controller will be able to counteract a particular disturbance, either through direct manipulation of the actuators or indirectly through loop tuning.

Then there's the problem of a potentially incomplete rule set. If a situation occurs that is not covered by the existing rules, the controller may not know what to do, and operators may have to intervene. Such an event would undoubtedly result in the addition of more rules, but even so the controller's performance would only be as good as the new rules and the skill of the experts who recorded them.

PICK ONE

So which adaptive controller is better? That remains to be seen. EXACT embodies the most classical techniques, so it could be considered the tried-and-true favorite. On the other hand, inexperienced users might prefer BrainWave or CyboCon if they're willing to simply trust that those controllers work as promised. Users with less faith might prefer INTUNE or KnowledgeScape so they can write or at least review the expert rules that the controller uses. Experienced process control users, on the other hand, might get the most out of Connoisseur.

And though it is tempting to think that any adaptive controller by its very nature should work equally well with any process, each has its forté. BrainWave, for example, is designed to be particularly effective with processes that demonstrate an appreciable delay or *deadtime* between the application of a control effort and its first effect on the process variable.

Similarly, every adaptive controller specializes in achieving a particular performance objective. BrainWave and EXACT are designed to drive the process variable toward the setpoint smoothly and rapidly without oscillations. INTUNE also tries to keep the process variable from overshooting the setpoint after a setpoint change, but gives the user a choice as to just how much overshoot is allowed—as little as possible, 10% to 20%, or no more than 35%.

CyboCon, on the other hand, attempts to minimize the accumulated squared error between the process variable and the setpoint (otherwise known as the *variance*). Connoisseur does the same while maintaining both the controller output and the process variable within their respective constraints.

Usability

Users will also find that some adaptive control products are simply easier to use than others. Some may prove particularly time consuming; some may require more technical expertise from the engineers who install them; and others may require more interaction with the operators who run them.

CyboCon and BrainWave, for example, are designed to be easy to use with minimal operator intervention. However, they both require the user to select a variety of operational parameters, some of which have more mathematical than physical significance. Fortunately, the correct choices for most of CyboCon's parameters are fairly intuitive. BrainWave's are somewhat less so. Connoisseur also requires the user to select several parameters, mostly for the benefit of the modeling operation. And though these would have some significance to a user who is already familiar with the behavior of his process, he would still need some instruction to set them correctly.

With some model-based techniques, users have considerable latitude to configure the model structure themselves and to personally supervise the identification operation. In fact, Connoisseur actually relies on the judgement and skill of the user, at least for initiating the modeling operation at the most opportune moment and terminating it once the model is "good enough." The user also has the option of breaking the process model into parts that can be updated independently. If some parts have already captured certain behaviors of the process well enough, there's no need to update the whole thing.

Whether having such options is an advantage or a disadvantage is a matter of the user's technical sophistication. If he already knows something about process modeling in general and the behavior of their process in particular, he can incorporate the known facts into the model so the modeling operation doesn't have to look so far and wide for an answer. He's also likely to recognize if the results are realistic or not. On the other hand, if he knows nothing about process modeling, then supervising the modeling operation could be a bewildering exercise, and the results may or may not make any sense to him.

Assumptions

Another question to consider when selecting a suitable adaptive control technique is the assumptions that are implicit in the controller's operations and what happens when the process does not meet them. For example, what does a pattern recognition controller do when it doesn't see any of the patterns it recognizes? What does a rule-based controller do when it has no rule for handling an unanticipated situation such as a process variable that ends up outside of its normal operating range?

For model-based techniques, perhaps the most critical assumptions are those that are implicit in the basic structure of the model, especially if it turns out that the assumed model can't be fit to the input/output data. A process can be so unusual (i.e., non-linear) that its behavior simply cannot be represented in the same form as the model, no matter how the parameters are set. And as previously noted, any modeling operation will fail if the input/output data is not sufficiently rich to adequately demonstrate the behavior of the process.

But even if the process is linear and driven by persistently exciting inputs, the modeling operation will produce inaccurate results in the presence of unknown or time-varying deadtime. Suppose, for example, that the controller assumes the last N outputs were the result of the last N inputs. If in fact those outputs were caused by inputs applied long ago (i.e., if the process has a long deadtime), the controller will end up identifying a model that may be mathematically correct, but useless for actually representing the behavior of the process.

There are three process characteristics in particular that a model-based controller must assume or glean from the operator before it can successfully generate a model and control the process:

- *Open-loop stability*. By default, most controllers assume that a finite change in the process input will have a finite effect on the process output. This is not the case for processes that are themselves unstable nor for processes like motors that integrate their inputs. The integrator problem is particularly easy to fix, but only if the controller knows about it up front.
- *Time frame*. No matter what curve-fitting technique it uses, a controller can only consider a finite collection of input and output samples as it attempts to identify the process model. It has to assume that all of the behavior it needs to see for modeling purposes is contained in that interval and that it is taking measurements fast enough to see it. If the process moves significantly faster or significantly slower than expected, essential data may be overlooked. This is undoubtedly why CyboCon requires the user to provide at least a rough estimate of the process's response time (though that does rather defeat the purpose of adaptive control, especially if the technique is supposed to be "model-free").
- *Inverse response*. It is generally a simple matter for the user to determine if a positive control action will have a positive or negative effect on the process variable and so inform the controller. However, processes that have a *nonminimum phase* or *inverse response* will react to a control action by first moving in the opposite direction before reversing course. This is a relatively rare phenomenon, so most controllers don't expect it. They generally assume that applying a control action to the process will always cause the process variable to move in the right direction, at least in the near term. A controller faced with an unexpected inverse response will start

backpedaling only to find that the process has already changed course on its own. This situation often leads to closed-loop instability.

OTHER CONSIDERATIONS

Technical functionality and user convenience are not the only criteria that could be considered when choosing an adaptive controller. History is another. Some products have a longer track record of successful applications than others. The original EXACT controller, for example, dates to the early 1980s. Connoisseur and INTUNE date to the mid-1980s while BrainWave and the present incarnation of EXACT date to the mid-1990s. CyboCon and KnowledgeScape are somewhat newer products.

Some adaptive controllers require more computing time than others. Rule-based techniques are generally faster than their model-based counterparts because their calculations are so much simpler. The difference in computational speed may or may not be appreciable, but it could limit the choice if the process is particularly fast.

The choice may also hinge on how the users think. Do the engineers already use mathematical models or expert rules for their own decision making? Do they think in terms of gain and phase margins, sensitivity functions, poles and zeros, or something else? Are the operators actively involved with running the process, or do they just leave matters up to the controllers they already have?

Then there are all the extra bells and whistles that differentiate one adaptive controller from another. Some might be just as important to a particular application as the controller's basic control functions. For example:

- EXACT, BrainWave, and CyboCon include adaptive feedforward compensators.
- CyboCon has various incarnations that are preconfigured for simple linear processes, deadtime-dominant processes, and nonlinear processes such as pH control.
- BrainWave, Connoisseur, and CyboCon can all be configured as multivariable controllers applicable to processes with several inputs and outputs.
- KnowledgeScape has an optimizer that can use the process model to search for setpoints that will best meet the user's objectives.
- Connoisseur offers several graphical utilities for manually massaging the input/output data.
- INTUNE's monitor mode organizes pertinent information about the controller's success (or lack thereof) in a convenient graphical format that helps users make maintenance and corrective decisions about not only the controller, but the process itself.

And finally, there's the issue of credibility. Is there any reason to believe that EXACT, Connoisseur, BrainWave, CyboCon, INTUNE, or KnowledgeScape is the superior product? Is there any evidence to support the claims that their developers have made? The following chapters should help answer those questions.

REFERENCES

Åström, K. J., and B. Wittenmark (1973). "On Self-Tuning Regulators," *Automatica* **9**, 185–199.

Kalman, R. E. (1958). "Design of a Self-Optimizing Control System," *ASME Trans.* **80**, 468–475.

Ziegler, J. G., and N. B. Nichols (1942). "Optimum Settings for Automatic Controllers," *ASME Trans.* **64**, 759–768.

1

ADAPTIVE TUNING METHODS OF THE FOXBORO I/A SYSTEM

Peter D. Hansen

This chapter will describe the controller structure and tuning strategy used to cope robustly with process behaviors such as unmeasured-load disturbances and variable deadtime (delay). The Foxboro I/A Series implements multivariable adaptive control using a technique called Exact MV. A single loop of Exact MV is implemented with PIDA, FBTUNE, and FFTUNE function blocks. The PIDA block is an advanced PID controller. The FBTUNE extender block performs adaptive tuning and gain scheduling of the PIDA feedback controller parameters. The FFTUNE extender block performs adaptation, gain scheduling, and feedforward compensation of measured-load and interacting-loop variables. Robust and rapid sampling-interval-independent methods are used for identifying and updating the process model upon which the adaptive controller tunings are based.

Linear proportional-integral-derivative (PID) feedback controllers are widely used in the process industries. Nevertheless, they are difficult to tune well. Process dynamics may be imperfectly known and may change over time. Production rate, feed composition, energy supply, and the environment affect behavior. Safety, inventory, and quality control loops often interact and are upset by measured and unmeasured loads. The processes, their measurements, and their manipulations are often nonlinear. This chapter will show how these difficulties can be overcome.

Additive or multiplicative feedforward compensation of a measured load can significantly reduce the load's effect on the controlled measurement provided there is no more delay in the manipulated variable's path to the controlled measurement than in the load's path. The compensation cannot have a negative delay. Because stability is not an issue, the compensator need only be effective at relatively low frequencies. Nevertheless feedforward compensation is rarely used. Process change can make the compensation ineffective. Manual retuning of a compensator may be difficult, requiring deliberate upsets of a possibly unmanipulatable load variable such as the weather. Adaptive gain scheduling of feedforward compensators based on responses to measured natural upsets overcomes these difficulties (Bristol and Hansen, 1987).

Feedforward compensation can also significantly reduce loop interactions. Decoupled feedback control pairs each manipulated variable with its principally affected controlled variable. For example, when two manipulated flows affect both inventory (liquid level) and quality (composition) variables, the relative gain array RGA (Bristol, 1966) shows that the larger flow should be used for feedback control of the inventory variable and the smaller for feedback control of the quality variable. Feedforward compensation can multiplicatively decouple the effect of the larger flow on the quality variable and/or additively decouple the smaller flow on the inventory variable (Figure 1.1).

A safety-variable controller may override a quality-variable controller when a safety threshold is approached. This can be done by selecting the smaller (or larger) of the two controller outputs to manipulate the process. Each controller output acts like

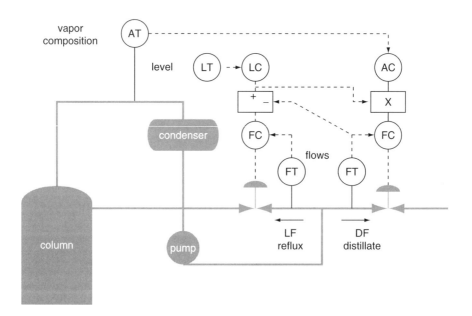

Figure 1.1   Local multivariable control.

the upper (or lower) limit of the other. Integral action in the unselected (limited) controller must not be allowed to wind up. An antiwindup strategy, which avoids overshoot on recovery from limiting when the controller is tuned for load rejection, freezes the controller's integral term when it is between the effective output limits and the output no longer affects the process. "Soft" limits may also be applied to a secondary controlled variable by limiting its setpoint, if its controller is tuned for unmeasured-load rejection (high-gain low-frequency feedback) and no overshoot to setpoint.

Measurement nonlinearity may be corrected with an external functional transformation such as square root for flow inferred from differential pressure or internal matched characterizations of setpoint and measurement for ion concentration inferred from pH. The effect of high-frequency measurement noise on the controller output can be reduced efficiently with a 0.7 damped quadratic (second-order Butterworth) low-pass filter. Its effective delay time is usually made a small fraction of the process effective delay in order not to degrade the ability to reject unmeasured load.

The effects of valve stick-slip friction, pressure-drop variation, and area versus stroke on a slow loop may be significantly reduced by containing the valve within a fast flow loop whose setpoint is manipulated by the slower outer loop. Functional transformation and pressure regulation may help to linearize the fast flow loop. A faster valve-positioning inner loop may help to reduce the effects of stick-slip friction. To prevent integral windup caused by valve rate limiting, the inner loop's integral time should be made at least as large as the time it would take for the valve to slew across its proportional band.

Process nonlinearity in quality-control loops may be reduced with multiplicative feedforward compensation. A manipulated flow is adjusted in proportion to a load flow. The outer-loop quality controller sets the ratio. Gain scheduling of the controller feedback and feedforward tuning parameters can also significantly improve performance by anticipating future behavior. Gain schedules are based on a process model together with a measured process load and/or the controller setpoint. These schedules can be adaptively tuned during normal operation to correct process-model mismatch.

There are many approaches to feedback controller design when a reliable linear process model is known (Panagopoulos et al., 1998). However, it is difficult to achieve a model good enough to accurately predict control loop stability limits, either analytically or experimentally. Good load rejection and stability depend critically on the loop effective delay (deadtime). Effective delay, which includes pure (transport) delay, nonminimum-phase zeros, and lags smaller than the two largest, is difficult to measure in the presence of dominant process lags, unmeasured load upsets, and measurement noise, particularly if deliberate large process upsets are not allowed.

This chapter will show how these difficulties have been overcome. Performance and robustness issues are addressed with an adaptive gain scheduling approach.

Controller structure for measured and unmeasured load rejection is discussed in the next section. Later, the structure of a minimum-variance feedback controller is compared with that of a PID. Although minimum-variance nominal performance is *ideal*, it is not robust with respect to process uncertainty or change. The subsequent section discusses a design approach intended to achieve robustness. It is shown to be intolerant of delay uncertainty if there can be high-frequency unity-gain crossings. Also, it yields poor unmeasured-load rejection when the process has a dominant lag.

Next, a direct algebraic PID controller design method is presented that achieves both good unmeasured-load rejection and target-tracking transient response shapes. However, a significant shift in a process time constant may degrade performance. The next section applies the algebraic tuning method to a controller with deadtime. The subsequent section presents the robust Exact MV method used in the Foxboro I/A Series for adapting gain scheduling to maintain feedback performance despite process change. Next is a discussion of feedforward control to reduce the effect of measured loads, followed by a presentation of the Exact MV method for adapting additive and multiplicative feedforward compensators. The final section presents some conclusions.

CONTROLLER STRUCTURE

The form of the controller is chosen to deal with manipulation, process, and sensor nonlinearities, and also process deadtime, process load structure, and system interactions and their changes over time. Nonlinearity compensations include matched setpoint and measurement input characterizers, an output correction for valve geometry and pressure drop, adaptive gain scheduling and multiplicative feedforward compensation for process nonlinearity. The basic PID controller includes a relative-gain-on-setpoint to allow tuning for both unmeasured-load rejection and nonovershooting setpoint response. Also, the controller may include a deadtime function in its integral feedback path to improve performance.

PID controllers have been used successfully to control a wide variety of processes. Proportional action is needed to stabilize very common dominant-lag processes. Integral action can stabilize a dominant-delay process and eliminate steady-state error caused by unmeasured or nonperfectly compensated loads. Derivative action is needed to stabilize a process with two dominant lags or to speed up the response of a process with one dominant lag.

Loads are typically applied upstream of dominant process lags. The initial part of the controlled variable's response to such an unmeasured load may be quite small, possibly resulting in an excessively small controller-output response unless the controller's feedback terms are tuned for load rejection. It is not sufficient to design for setpoint response, since good setpoint tracking can be achieved with feedforward (open-loop) terms alone. The controller structure should allow independent tuning of feedback and feedforward terms since both good load rejection and setpoint tracking are important objectives.

Process deadtime (delay) plays a crucial limiting role in determining an optimally tuned loop's response to an unmeasured-load upset. The optimal integrated absolute error (IAE) response to an unmeasured load step is proportional to deadtime for a pure deadtime process, to deadtime squared for a dominant-lag process, and to deadtime cubed for a dominant-double-lag process. Since a digital controller's measurement-sampling interval and output-update interval contribute to the open-loop effective delay, it is particularly important for achieving good unmeasured-load rejection to choose these intervals to be small relative to the process delay and not size them relative to the open-loop settling time.

Figure 1.2 is the Bode amplitude versus frequency plot for two PID controller tuning strategies with a dominant-lag process. The dotted curve is the inverse process. The solid curve is the controller. The logarithm of the open-loop absolute gain is the vertical difference between the controller and inverse process curves. The open-loop gain is 1 where the curves intersect. Plotting the Bode diagram in this way helps to visualize independently the effects on stability of controller and process changes. A stable loop has a net slope of nearly −1 at the lowest frequency crossing. Here the process effective delay (including nonminimum phase zeros and lags smaller than the largest two) should contribute less than 90° of phase lag.

For a dominant-lag process the intersection occurs in the central region of the controller curve where the proportional term is dominant. Note that when the controller is tuned for load rejection, the open-loop gain is much higher at low frequency.

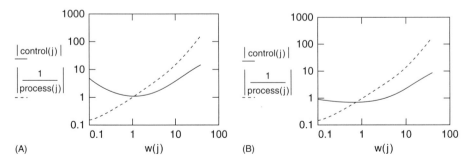

Figure 1.2 *(A) Tuned for load rejection and (B) tuned for setpoint tracking.*

If the process were a dominant delay, the intersection would occur in the low-frequency region of the controller curve where integral action is dominant. If the process were a pure delay, its gain curve would be flat. To avoid second and third intersections and potential high-frequency instability, derivative action should not be used. If the process has two dominant lags, the intersection will occur in the high-frequency section of the controller curve where derivative action is dominant.

In each of these example loops, the net slope is approximately 1 at the intersection, causing the net phase lag there to be between 90 and 180°. Therefore, the PID controller can be tuned to accommodate a process that appears to be a lag-lag-delay in the neighborhood of the unity-gain intersection. The PID controller may also be effective on a process with a dominant resonant quadratic since the process behaves like a dominant double-lag process near the unity-gain intersection.

Unfiltered derivative action may produce very high controller gain at the Nyquist frequency, the highest meaningful frequency in a digital control loop. A digital low-pass measurement filter may be used to reduce high-frequency process and measurement noise amplification and to prevent the open-loop gain from crossing unity more than once. A first-order filter limits the controller high-frequency gain. A second-order Butterworth filter is more effective because it "rolls off" the useless high-frequency gain without necessarily contributing more low-frequency phase lag or effective delay.

Second and higher derivative actions are not included in general-purpose controllers. These terms would amplify high-frequency noise, causing excessive controller-output action. Also, useful tuning of these actions would require knowledge of the difficult-to-measure high-frequency input/output behavior of the process. Second and higher integral actions are also not included in general-purpose controllers. These terms would excessively slow the closed-loop response and not improve the rejection of an unmeasured-load step. Cascade PID control structures, which usually require measurement and control of additional "state" variables, are more effective in achieving the benefits that higher-order integral and derivative terms might bring.

Integral action may be achieved with a first-order lag in a positive-feedback path.

$$u = \varepsilon + \frac{1}{1 + Is} u = \left(\frac{1}{Is} + 1\right) \varepsilon \qquad (1.1a)$$

If a delay is included with the lag, a different and more effective kind of integral action is achieved.

$$u = \varepsilon + \frac{e^{-\tau s}}{1 + Is} u = \left(\frac{1 + Is}{Is + 1 - e^{-\tau s}}\right) \varepsilon \qquad (1.1b)$$

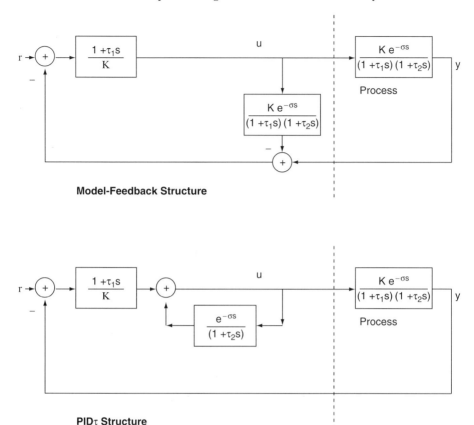

Figure 1.3 Equivalent model-feedback and PIDτ controller structures.

At low frequencies ($s = j\omega$) the effective integral time is $I + \tau$. A PID with this type of integration is called a PIDτ or deadtime controller. Shinskey (1994) shows that at high frequencies this operator with large τ adds phase lead, thus enabling tighter control than with conventional integral action. A model-feedback control structure for a lag-lag-delay process is shown in Figure 1.3 to be equivalent to a PIDτ structure.

This PIDτ structure is said to be "interacting" since its transfer function from measurement to output is a product of factors. Its numerator has two real zeros. In a noninteracting controller, the transfer function is a sum of terms. Its numerator may have complex-conjugate or real zeros. For unmeasured-load rejection, complex zeros are useful when the process has a secondary lag larger than its effective delay.

Another important structural distinction is whether the same proportional and derivative actions are applied to setpoint as are applied to measurement. The integral term must be applied to setpoint-minus-measurement in order to eliminate steady-state modeling error. However, the measurement gain and derivative terms can be tuned for unmeasured-load rejection and the setpoint-gain term adjusted to avoid excessive overshoot to a setpoint step.

Table 1.1 shows the various PID and PIDτ controller structures. Where the noninteracting PID numerator has an *I* in its middle term, the PIDτ has a *D*. This enables maximum use of the delay-type integration in the PIDτ, while still achieving a quadratic numerator with only one differentiation.

A useful performance measure is the integrated control error *IE* in moving from one steady state to another. This measure can be calculated from the controller tuning constants and the changes in setpoint and controller output.

Table 1.1 Controller Structures

	Design for Unmeasured-Load Rejection	
	Interacting	Noninteracting
PID	$u = \frac{(1+AIs)r - (1+Is)(1+Ds)y_f}{PIs}$	$u = \frac{(1+AIs)r - (1+Is+IDs^2)y_f}{PIs}$
PIDτ	$u = \frac{(1+Is)(r - (1+Ds)y_f)}{P(Is+1-e^{-\tau s})}$	$u = \frac{r - (1+Ds+Is^2)y_f}{P(Is+1-e^{-\tau s})}$

	Design for Setpoint Tracking (Pole Cancellation)	
	Interacting	Noninteracting
PID	$u = \frac{(1+Is)(1+Ds)(r-y_f)}{PIs}$	$u = \frac{(1+Is+IDs^2)(r-y_f)}{PIs}$
PIDτ	$u = \frac{(1+Is)(1+Ds)(r-y_f)}{P(Is+1-e^{-\tau s})}$	$u = \frac{(1+Ds+IDs^2)(r-y_f)}{P(Is+1-e^{-\tau s})}$

s is the differential operator, $d(\ )/dt$
u is the controller output
r is the setpoint
y_f is the filtered controlled variable

$$y_f = \frac{1}{1 + \tau_f s + .5(\tau_f s)^2} y \tag{1.2}$$

y is the controlled measurement
τ_f is the measurement filter time constant
P is the proportional band (inverse gain)
I is the integral time
D is the derivative time
A is the relative gain applied to setpoint, $0 \leq A \leq 1$
τ is the delay time

$$IE = \int_{t_1}^{t_2}(r-y)dt = \int_{t_1}^{t_2}(r-y_f)dt - \tau_f \delta r = (I+\tau)(P\delta u + (1-A)\delta r) - \tau_f \delta r \quad (1.3a)$$

If the response is nonovershooting, IE is equal to the integrated absolute error IAE.

For a load change, δu is independent of tuning. Best load rejection is achieved when the value of $(I + \tau)P$ is smallest, consistent with a nonovershooting response to a load step applied at the controller output. Process loads are usually applied upstream of dominant process lags and can be approximated as a sequence of steps, a single step at the controller output being a severe test.

For a setpoint change $\delta u = \delta r/K$, where K is the process gain,

$$IE = (I+\tau)\left(\frac{P}{K} + 1 - A\right)\delta r - \tau_f \delta r \quad (1.3b)$$

A dominant-lag loop, with controller tuned for unmeasured load rejection, has a small P/K ratio. If A were 1, the setpoint IE would be nearly zero, much smaller than the IAE. The error response to setpoint would have significant overshoot with nearly equal area above and below zero. Since P cannot be increased without hurting unmeasured-load rejection, A must be made substantially less than 1 in order to prevent large setpoint overshoot.

In the next sections minimum variance (and LQG), $\|H\|_\infty$ (pole cancellation), and algebraic tuning are described and compared for performance and robustness. The concept of frequency margin is introduced.

MINIMUM VARIANCE CONTROL

Ideal controllers for several *m*-integral-delay processes are presented next. These linear processes are salient extremes of common process models, dominant delay (deadtime), dominant lag, and dominant quadratic lag. A load in a real process is usually introduced upstream of dominant process lags and has a nonzero mean value. Therefore, the *ideal* controller rejects a step (or random-walk) load added to the controller output by returning the controlled variable to its setpoint value one delay time after its upset is first sensed. Also required is an *ideal* controlled variable response to a setpoint step, a pure delay equal to the process delay. Measurement noise is considered nonexistent, controller action is not constrained, and robustness is not an objective. Therefore this *ideal* control may not be practical. It is a form of minimum-variance control, an extreme form of linear–quadratic–Gaussian (LQG) control.

The *m*-integral-delay process is

$$y = \frac{e^{-\tau s}}{(Ts)^m}(u+v) \quad (1.4a)$$

where y is the controlled measurement, u is the controller output, and v is the unmeasured load input, T is the process integral time, τ is the process delay time, and s is the time-derivative operator.

The *ideal* closed-loop transfer relation is

$$y = e^{-\tau s} r + \frac{e^{-\tau s}}{(Ts)^m}\left(1 - e^{-\tau s}\sum_{i=0}^{m}\frac{(\tau s)^i}{i!}\right)v \tag{1.4b}$$

This is achieved with the following controller, where r is the setpoint:

$$u = \frac{r - \left(\sum_{i=0}^{m}\frac{(\tau s)^i}{i!}\right)y}{\frac{1}{(Ts)^m}\left(1 - e^{-\tau s}\sum_{i=0}^{m}\frac{(\tau s)^i}{i!}\right)} \tag{1.4c}$$

The first nonzero term of the denominator Taylor series $\frac{\tau^{m+1}s}{(m+1)!T^m}$ reveals its low-frequency integrating behavior. A PID controller, with no measurement filter and no proportional gain or derivative applied to the setpoint, has a similar numerator form for $m < 3$, but its denominator integrating action is achieved very differently.

The ($m = 0$) unity-gain pure-delay inverse process and its *ideal* controller are shown on the Bode plot of Figure 1.4A. The controller's first cusp has a shape something like that of a PID controller. This style of integral action is exploited in the PIDτ controller where it is applied to processes with lags as well as delay (Shinskey, 1994;

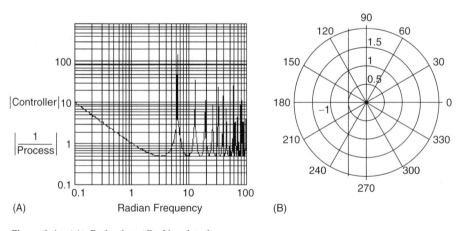

Figure 1.4 (A) Bode plot; (B) Nyquist plot.

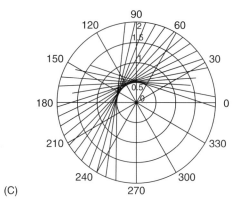

(C)

Figure 1.4 (C) Nyquist plot with 2% delay mismatch.

Hansen, 1998). Figure 1.4B is a polar version of the Nyquist plot. The trajectory makes an infinite number of clockwise encirclements of the origin and none of the −1 point. Thus the loop is stable. The gain margin is 6 dB (factor of 2) and the phase margin is 60°, apparently indicating adequate robustness. However these traditional stability margins are misleading. The frequency (or delay) margin is 0 dB (factor of 1). The slightest mismatch in either direction between the process and controller delays causes instability at a high frequency where the mismatch contributes at least $60° (\omega|\delta\tau| \geq \pi/3)$. This is illustrated in Figure 1.4C where the process delay is 2% larger than the controller delay.

The inverse integral-delay ($m = 1$) process and its controller are shown on the Bode plot of Figure 1.5A. The polar Nyquist plot for the integral-delay process is shown in Figure 1.5B. In the neighborhood of the first crossing the controller behavior is like that of a PID. The trajectory makes an infinite number of counterclockwise encirclements of the −1 point. Although the loop is stable, it is not robust. A small change in the high-frequency behavior of the process could destabilize the loop. Bode and Nyquist plots for m greater than 1 are similar.

LQG control improves loop robustness for a self-regulating process by penalizing controller output changes as well control errors. However, increasing the relative penalty on output changes may reduce robustness of a loop that requires an active feedback controller to achieve stability.

CONTROL BY MINIMIZING SENSITIVITY TO PROCESS UNCERTAINTY

Good control performance requires the controller tuning to be performed with a less than perfect process characterization. Phase information in the frequency neighborhood of the inverse process and controller absolute gain crossings is critical for predicting feedback-loop stability margins. Effective delay, which contributes

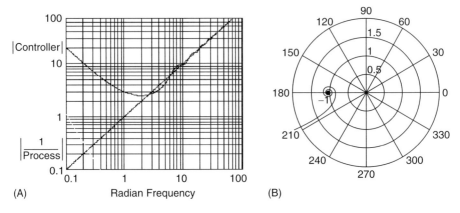

Figure 1.5 (A) Bode plot; (B) Nyquist plot.

significantly to phase at these frequencies, may be obscured from accurate open-loop measurement or calculation by dominant process lags and noise. Zero-frequency gain, as distinct from gains in the critical frequency regions, is most easily measured or calculated but may have little impact on stability.

A linear loop (represented in the complex frequency domain) with controller c and process $g(1+\delta)$ has $1+cg(1+\delta)=0$ at the stability limit. Here g is the nominal process and $1+\delta$ is the factor that would bring the loop to the stability limit. In terms of the nominal loop function

$$h = (1+cg)^{-1}cg \tag{1.5a}$$

$$1 + h\delta = 0 \tag{1.5b}$$

at the stability limit. The distance to a point on the Nyquist trajectory from the origin is $|cg|$ and from the -1 point is $|1+cg|$. Thus in Figure 1.4B, where all points on the Nyquist trajectory are equidistant from the origin and the -1 point, $|h|=1$.

If the product of the largest $|\delta|$ over all frequencies $\|\delta\|_\infty$ with the largest $|h|$ over all frequencies $\|h\|_\infty$ is less than 1, the process must be stable. Therefore the norm $\|h\|_\infty$ is used as a measure of robustness: the smaller its value, the larger $\|\delta\|_\infty$ may be. However, if there can be only one unity-gain crossing, this measure may be too conservative. If the process uncertainty factor $1+\delta$ is a unity-gain delay, $\|\delta\|_\infty = 2$. This robustness criterion would require $\|h\|_\infty < 0.5$ and eliminate from consideration a controller or process with an integrating factor, for which $|h|=1$ at zero frequency. Even a process uncertainty factor $1+\delta$ that approaches zero at high frequencies has norm $\|\delta\|_\infty \geq 1$, apparently restricting the nominal loop norm $\|h\|_\infty < 1$.

A control-loop design strategy, where the controller provides integral action and minimizes $\|h\|_\infty$ makes $\|h\|_\infty = 1$ and provides a 60° phase margin (but possibly no delay margin). This can be achieved, provided the open-loop process is stable, with a controller that cancels dominant process poles and minimum-phase zeros, while structuring and tuning its integral action to cope with the process effective delay. For example, the PID controller c for a process g with a gain k, delay τ_d, and two lags τ_1, τ_2, each greater than $\tau_d/3$, would use its quadratic numerator to cancel the two process lags and deal with the gain and delay with its denominator $PI = 2(k\tau_d)_{\max}$. The $P\,I$ product is equal to the integrated control error for a unit load change. The controlled variable y and controller output u responses to load v and setpoint r, assuming all three P, I, D terms are applied to the control error and that the controller output does not limit, are

$$y = g(u+v) = h\left(r + \frac{v}{c}\right), \quad u = c(r-y) = h\left(\frac{r}{g} - v\right) \tag{1.6a}$$

$$g = \frac{ke^{-\tau_d s}}{(1+\tau_1 s)(1+\tau_2 s)}, \quad c = \frac{(1+\tau_1 s)(1+\tau_2 s)}{PIs} \tag{1.6b}$$

$$h = \frac{gc}{1+gc} = \frac{1}{1 + \frac{(k\tau_d)_{\max}}{k\tau_d}\left(2\tau_d s + 2(\tau_d s)^2 + (\tau_d s)^3 + \ldots\right)} \tag{1.6c}$$

Although this approach is very tolerant of changes in the $k\tau_d$ product, it has several drawbacks when a process lag is large compared with the delay. The controller responds much more aggressively to setpoint changes and measurement noise than to load changes applied upstream of the process lag, causing large controller output sensitivity to measurement noise and poor unmeasured-load rejection. Also, response to a setpoint step may be slower than expected because weak low-frequency feedback fails to keep the controller output at its limit until the measurement nearly reaches its setpoint (Hansen, 1998). Furthermore, this approach cannot be used on a nonself-regulating process. An integral-delay or double-integral-delay process with a controller having integral action and tuned for load rejection may have to tolerate $\|h\|_\infty$ greater than 2 and a phase margin less than 30°.

ALGEBRAIC CONTROLLER DESIGN FOR LOAD REJECTION AND SHAPED TRANSIENT RESPONSE

The Exact MV preferred approach to PID controller design uses an additive compensation in the s domain to shape the closed-loop transfer function rather than the multiplicative compensation of the pole-cancellation approach. The method can be applied to a nonself-regulating, statically unstable, or lightly damped resonant process containing time delay.

Tuning for load rejection can be significantly simplified by expressing the inverse-process transfer function as a Taylor series in $s = j\omega$ convergent in the neighborhood of the first potential open-loop unity-gain and 180° phase crossings:

$$(a_0 + a_1 s + a_2 s^2 + a_3 s^3 + \ldots) y_f = u + v \tag{1.7a}$$

where v the unmeasured load, y_f the controlled filtered measurement, u the controller output, and a_0 are nondimensional. The denominator series for a large positive process lead may not be convergent in the frequency range critical for stability. In this case a large positive process lead should be paired with a large process lag and their high-frequency ratio treated as a gain:

$$\frac{1 + \text{leads}}{1 + \text{lags}} \approx \frac{\text{lead}}{\text{lag}} \tag{1.7b}$$

The denominator series for a time delay is

$$e^{\tau s} = 1 + \ldots + \frac{1}{n!}(\tau s)^n + \ldots \tag{1.7c}$$

The controller measurement filter is a 0.7 damped low-pass filter whose time constant is usually small (0.025) compared with the process delay. It is treated as part of the process.

The series expansion for a gain-lag-lag-delay (K, τ_1, τ_2, τ_d) process becomes

$$\begin{aligned}
a_0 K &= 1 \\
a_1 K &= \tau_1 + \tau_2 + \tau_d \\
a_2 K &= \tau_1 \tau_2 + (\tau_1 + \tau_2)\tau_d + \frac{\tau_d^2}{2} \\
a_3 K &= \tau_1 \tau_2 \tau_d + (\tau_1 + \tau_2)\frac{\tau_d^2}{2} + \frac{\tau_d^3}{6} \\
&\vdots
\end{aligned} \tag{1.7d}$$

The PID controller is designed in two steps to ensure that derivative action is not used on an all-pass process such as a pure delay. The first step is an inner-loop PD controller:

$$u = k_s r_i - (k_m + d_m s) y \tag{1.8a}$$

Derivative d_m action is not applied to the inner-loop setpoint r_i, and the gain k_s applied to setpoint may not equal k_m applied to measurement y. Eliminating u gives the closed inner loop:

$$(k_m + a_0 + (d_m + a_1)s + a_2s^2 + a_3s^3 + \ldots)y = k_s r_i + v \quad (1.8b)$$

Note that the controller parameters k_m and d_m affect only the lowest-order closed-loop terms that dominate the low-frequency behavior. Stable closed-loop behavior can be guaranteed if the open-loop absolute gain crosses unity only once. Maximum use of feedback results from tuning the closed inner loop to behave like a delay at low frequencies. Thus inner-loop feedback will not be used if the open-loop process is already a delay. The target behavior is

$$e^{\tau_i s} y = \left(1 + \tau_i s + \frac{(\tau_i s)^2}{2} + \frac{(\tau_i s)^3}{6} + \ldots\right) y = r_i + \frac{v}{k_s} \quad (1.8c)$$

Since there are four unknowns, terms through s^3 are matched. It is assumed that higher degree terms of the process (combined with the quadratic measurement filter) are well behaved, causing no hidden open-loop unity crossings. Equating term by term:

$$\tau_i = \frac{3a_3}{a_2}, \quad k_s = \frac{2a_2}{\tau_i^2}, \quad d_m = k_s \tau_i - a_1, \quad k_m = k_s - a_0 \quad (1.8d)$$

If the process were a pure delay, k_m and d_m would be 0.

For a valid result, τ_i must be greater than 0. If τ_i is not positive or if less than the calculated derivative action is required, and a_1 is not 0 and has the same sign as a_2, a_3 may be artificially assigned:

$$a_3 \Leftarrow \frac{2a_2^2}{3a_1} + \left(a_3 - \frac{2a_2^2}{3a_1}\right) df. \quad (1.8e)$$

If the detuning factor, $0 \leq df \leq 1$, is set to 0, the derivative term d_m becomes 0 and τ_i becomes $2a_2/a_1$. Alternatively, when the calculated d_m is negative and the open-loop process may have a lightly damped high-frequency resonance, the 0.7-damped quadratic (Butterworth) measurement filter,

$$y_f = \frac{y}{1 + \tau_f s + .5(\tau_f s)^2} \quad (1.8f)$$

can be positioned to make $d_m = 0$. This requires solving a quartic (a quadratic if a_0 is 0) for the smallest positive real root.

$$(3a_1 a_3 - 2a_2^2) + (3a_0 a_3 - a_1 a_2)\tau_f + \left(a_0 a_2 - \frac{a_1^2}{2}\right)\tau_f^2 - \frac{a_0 a_1}{2}\tau_f^3 - \frac{a_0^2}{2}\tau_f^4 = 0 \quad (1.8g)$$

Alternatively, when the calculated k_m is negative, high-frequency instability is unlikely, or the quartic has no positive real root, d_m may be made zero by replacing a_3 with $\frac{2a_2^2}{3a_1}$. In either case if the recalculated k_m is negative, the time constant τ_f of a first order measurement filter should be positioned to make k_m zero,

$$\tau_f^2 = 2\frac{a_2}{a_0} - \left(\frac{a_1}{a_0}\right)^2 \tag{1.8h}$$

and the controller should use only integral action:

$$u = \frac{a_0^2(r-y)}{2.5(a_1 + a_0\tau_f)s} \tag{1.8i}$$

Normally the outer-loop controller is a PI type,

$$r_i = \left(\frac{1}{i_0 s} + k_0\right)(r - y) \tag{1.9a}$$

tuned to control the closed inner loop:

$$y = e^{-\tau_i s}\left(r_i + \frac{v}{k_s}\right) \tag{1.9b}$$

The closed outer loop becomes

$$\left(\frac{1}{i_0 s} + (k_0 + 1) + \tau_i s + \frac{(\tau_i s)^2}{2} + \frac{(\tau_i s)^3}{6} + \dots\right) y = \frac{(1 + k_0 i_0 s) r}{i_0 s} + \frac{v}{k_s} \tag{1.9c}$$

A well-shaped target function is a cascade of n equal lags:

$$\left(1 + \frac{\tau_0 s}{n}\right)^n y = (1 + k_0 i_0 s) r + \frac{i_0 s}{k_s} v =$$
$$\left(1 + \tau_0 s + \frac{n-1}{2n}(\tau_0 s)^2 + \frac{n-1}{2n}\frac{n-2}{3n}(\tau_0 s)^3 + \frac{n-1}{2n}\frac{n-2}{3n}\frac{n-3}{4n}(\tau_0 s)^4 + \dots\right) y \tag{1.9d}$$

This causes the load step response to approximate a Gaussian probability-density function and the setpoint step response to approximate its integral. Solving term by term gives

$$n = 11, \quad \tau_0 = \frac{11}{6}\tau_i, \quad i_0 = \frac{55}{36}\tau_i, \quad k_0 = \frac{1}{5} \tag{1.9e}$$

The PID controller has the form:

$$u = \frac{1}{P}\left(\left(\frac{1}{Is} + A\right)r - \left(\frac{1}{Is} + 1 + Ds\right)y_f\right) \tag{1.10a}$$

where

$$P = \frac{1}{k_0 k_s + k_m}, \quad I = \frac{i_0}{k_s P}, \quad D = d_m P, \quad A = k_0 k_s P \tag{1.10b}$$

Table 1.2 applies this (Equation 1.10b) tuning procedure to the m-integral-delay processes. Also presented is the integrated error IE for a unit load step.

Bode and Nyquist plots for these loops are shown in Figures 1.6, 1.7, and 1.8. The log of the gain factors k_{max} and k_{min} indicate how much the inverse process Bode plot can be shifted down and up relative to the controller before a stability limit is reached. The log of frequency factors f_{max} and f_{min} indicate how much the inverse process Bode plot can be shifted left and right relative to the controller before a stability limit is reached. The quadratic measurement filter attenuates the controller at high frequencies, helping to avoid a high-frequency instability that might be caused by a neglected process lead or resonant lag.

Setpoint and load step responses are shown in Figure 1.9. The controller output response is not overly aggressive, but comparably aggressive for both setpoint and unmeasured load steps.

ALGEBRAIC TUNING OF A CONTROLLER WITH DEADTIME

Algebraic tuning of the noninteracting PIDτ

$$u = \frac{r - (1 + Ds + IDs^2)y_f}{P(Is + 1 - e^{-\tau s})} \tag{1.11a}$$

for unmeasured-load rejection is executed in a single step to make the closed loop

$$(1 + Ds + IDs^2 + P(Is + 1 - e^{-\tau s})(a_0 + a_1 s + a_2 s^2 + a_3 s^3 + \ldots))y_f \tag{1.11b}$$
$$= r + P(Is + 1 - e^{-\tau s})v$$

Table 1.2 Tuning for Performance ($df \geq 1$)

Process	a_0	a_1	a_2	a_3	P	I	D	A	IE	k_{min}	k_{max}	f_{min}	f_{max}
Delay	1	τ	$\tau^2/2$	$\tau^3/6$	5.0	0.3τ	0.0	1.0	1.5τ	0.0	2.8	0.0	2.7
Integral-delay	0	T	$T\tau$	$T\tau^2/2$	$0.938\tau/T$	2.7τ	0.313τ	0.167	$2.53\tau^2/T$	0.0	1.8	0.4	2.0
2Integral-delay	0	0	T^2	$T^2\tau$	$3.75\tau^2/T^2$	5.4τ	2.5τ	0.167	$20.3\tau^3/T^2$	0.4	1.8	0.5	1.3

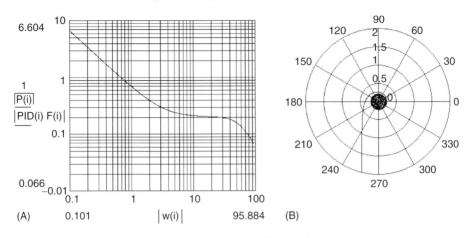

Figure 1.6   Pure delay process: (A) Bode plot; (B) Nyquist plot.

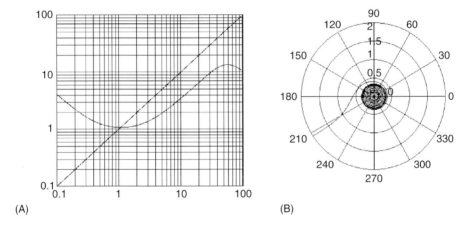

Figure 1.7   Integral-delay process: (A) Bode plot; (B) Nyquist plot.

approximate a multiple $(1/m)$ equal-lag target:

$$(1 + mT_0 s)^{\overset{\perp}{m}} y_f = r + P(Is + 1 - e^{-\tau s})v \qquad (1.11c)$$

There are six unknowns (P, I, D, τ, T_0, and m) that are found with some difficulty by matching terms through sixth order in the series expansions. If the process model is accurate, this tuning results in nonovershooting responses, closely resembling Gaussian probability distributions, to load and setpoint steps. If the process were a pure delay $K \cdot e^{-\sigma \cdot s}$, the tuning calculation would yield $m = 0$ and the ideal model-feedback controller ($\tau = T_0 = \sigma$, $P = K$, $I = D = 0$).

42   Techniques for Adaptive Control

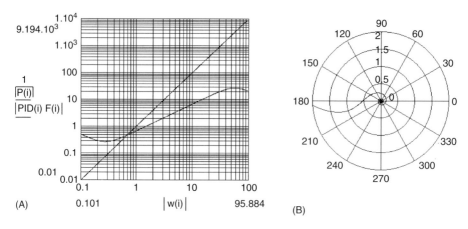

Figure 1.8 2Integral-delay process: (A) Bode plot; (B) Nyquist plot.

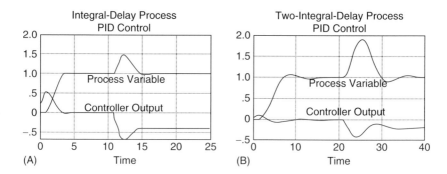

Figure 1.9 Setpoint and load responses.

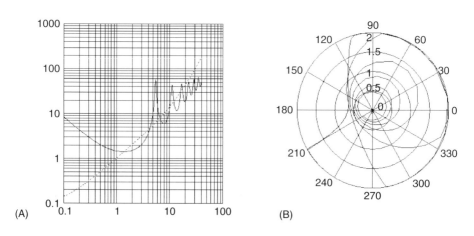

Figure 1.10 PIDτ control of a dominant-lag process: (A) Bode plot; (B) Nyquist plot.

Because of the multiple cusps in the controller Bode amplitude curve, it is very likely that there will be multiple open-loop unity-gain crossings as shown in Figure 1.10A for an integral-delay process. To achieve stability, the open-loop phase at each of these intersections must be well behaved as shown in the corresponding Nyquist diagram of Figure 1.10B. In this figure absolute gains greater than 2 were clamped at 2. Note that the -1 point (radius $= 1$, angle $= 180°$) is on the left of the increasing-frequency trajectory, indicating a stable closed loop. A change in process delay without a corresponding adjustment of controller delay might cause a closed-loop high-frequency instability. The destabilizing effect of process-and-controller-delay mismatch can be reduced by low-pass filtering that eliminates some of the high-frequency open-loop unity-gain crossings.

A process with one or two dominant lags is difficult to identify from input/output measurements with sufficient accuracy to make reliable tuning calculations for good closed-loop performance. Noise tends to obscure the high-frequency behavior where the effects of delay become significant. The controller is difficult to tune manually for load rejection, because near optimal tuning, both increasing and decreasing the proportional band P can make the loop less stable. Table 1.3 shows tuning and performance for several tuning strategies.

In the first column of Table 1.3, controller types are listed. The upper four are tuned for load rejection. The PIDi is the interacting PID. In the last two rows the controllers are designed for setpoint response. The P.C. (pole cancellation) is an interacting PID where the poles of the process are cancelled with the zeros of the controller. The M.F. is a model-feedback controller similar to the interactive PIDτ, except that it uses a second derivative term, listed in the I column, instead of a true I term, to create controller zeros to cancel the process poles. The next five columns are tuning constants.

The final column is the IE for a unit load step added to the controller output $P(I + \tau)$, which is also the IAE, since the tuning achieves small or no overshoot. Proportional plus integral PI control is inappropriate for this process, hence its very large IAE. The IAE of the PIDτ is half that of the noninteracting PID, which itself is

Table 1.3 Tuning and Performance for Various Controllers for a Process with Two Equal Lags 10 Times Larger Than the Unit Delay

	P	I	D	A	τ	IAE
PI	0.8362	11.2498	0.0000	0.3060	0.0000	9.4074
PID	0.0267	4.8085	1.7858	0.1711	0.0000	0.1284
PIDi	0.0534	3.5716	3.5716	0.3422	0.0000	0.1908
PIDτ	0.0225	1.4429	3.8345	0.0000	1.2791	0.0612
P.C.	0.2000	10.0000	10.0000	1.0000	0.0000	2.0000
M.F.	1.0000	10.0000	10.0000	1.0000	1.0000	1.0000

two-thirds that of the interacting PIDi. Although the *IAE* of the model-feedback is half that of the pole-cancellation controller, it is more than 8 times greater than that of the PID. Load-and-setpoint-step responses for four of these controllers are shown in Figure 1.11.

If the pole-cancellation controller output is limited to the range of the noninteracting PID, its setpoint response degrades substantially. This is also true for the model-feedback controller unless the controller is "redesigned" each time step with a constrained-output and measurement trajectory optimization.

When the PID or PIDτ output limits, its integral term is frozen. The high proportional and derivative gains maintain fast recovery with small overshoot.

ROBUST ADAPTATION OF FEEDBACK CONTROLLER GAIN SCHEDULING

Adaptive tuning, responsive to actual closed-loop behavior, may be the most reliable method of achieving both good performance and robustness. Single-step convergence is achieved by calculating new tunings from an updated process model. A linear differential-equation model with explicit time delay allows arbitrarily small sampling and control intervals. Thus, sampling and control delays need not add significantly to the open-loop delay. As has been shown, delay (deadtime) has an adverse effect on unmeasured load rejection.

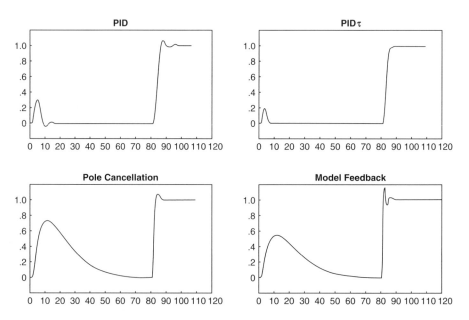

Figure 1.11 Load-and-setpoint-step responses, dominant lag-lag process.

Adaptive Tuning Methods of the Foxboro I/A System

A pretune method is required to achieve the initial process model and controller tuning. While the process is steady and open loop, a doublet pulse is applied at the controller output and the measured variable is monitored. The pulse height is user determined. The pulse width is determined by the measurement crossing a user-set change threshold. This causes the pulse width to be slightly larger than the process delay. The doublet pulse causes minimal upset to the process measurement while providing useful information in the frequency range critical for closed-loop stability (Figure 1.12). In the I/A Series Exact MV implementation, response-pattern features are used to determine a 1 or 2 lag-gain-delay model of the process.

The gain-lag-delay model is particularly easy to identify from response pattern features. The algebraic-tuning method is used to calculate the initial controller tuning:

$$Gain = -\frac{y_1^2}{y_2 b} \qquad (1.12a)$$

$$Lag = -\frac{T}{\ln\left(1 + \frac{y_2}{y_1}\right)} \qquad (1.12b)$$

$$Delay = t_1 - T \qquad (1.12c)$$

The feedback tuner also uses a self-tune method to update tuning based on response pattern features arising from normally occurring (possibly unmeasured) upsets while the control loop is closed. When an oscillatory response is detected, the gain and frequency shifts of the process relative to the controller are determined and used to shift the controller to restore desired performance and robustness.

Robust recovery from an oscillatory condition requires a reliable model update when only a conjugate pair of lightly damped closed-loop roots is excited. In this case two process parameters can be updated, the process factors for gain k and frequency (or time scale) f. The autonomous closed-loop complex equation

$$c\{s\} + \frac{1}{kg\{fs\}} = 0 \qquad (1.13a)$$

Figure 1.12 Pretune doublet-pulse response.

can be expressed as two real equations, for phase and amplitude. First the phase equation is solved iteratively for the frequency factor f. Then the amplitude equation is solved for the gain factor k. The solution selected has positive values of f and k close to 1. The controller $c\{s\}$ and the process-model $g\{s\}$ transfer functions are known from the previous update. The complex frequency

$$s = \omega(-\alpha \pm j) \qquad (1.13b)$$

is calculated from measurements of four successive controller error or output peak values x_i and times t_i. Redundancy is used to isolate the oscillatory mode:

$$e^{-2\pi\alpha} \approx \min\left(\left(\frac{-x_2 + 2x_3 - x_4}{x_1 - 2x_2 + x_3}\right)^2, \frac{x_3 - x_4}{x_1 - x_2}\right), \quad \omega \approx \frac{2\pi}{t_3 - t_1} \qquad (1.13c)$$

The search for peaks starts when the error or output change exceeds its user-set threshold (Figure 1.13).

The Bode plot of the inverse process is assumed to translate vertically and horizontally relative to the controller in the frequency neighborhood of the oscillation. Retuning translates the Bode plot of the controller similarly to nearly restore the original gain and frequency margins. If the complex frequency is imperfectly identified or if there is process-model Bode-plot shape mismatch, an additional fine retuning may occur after the next upset using this tuning set.

Expert rules are used when the control-error response is not oscillatory or when oscillations are limit cycles or externally forced.

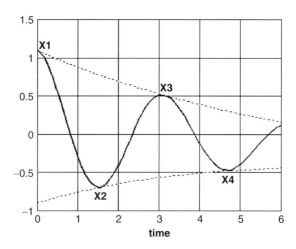

Figure 1.13   Self-tune peak search.

Adaptive Tuning Methods of the Foxboro I/A System   47

This self-tuning method can be used for any single-loop feedback controller structure (e.g., the PID or PIDτ) commissioned with any tuning strategy (e.g., algebraic or pole cancellation).

The Exact MV pretune and self-tune functions are demonstrated with an I/A Series control processor. A simulated lag-delay process is controlled by a PIDA controller in Figure 1.14. Initially the lag and delay were equal, both 12 seconds. With the controller in *manual* (open loop) and the process measurement steady at 50% of span, FBTUNE's pretune was run. A 10% amplitude doublet pulse was applied at the controller output (the lower curve). As soon as pretune identified the process and established the initial controller tuning, the controller was automatically switched to *auto* (closed loop) and the measurement moved without overshoot to the 70% setpoint. Then the simulated process delay was doubled to 24 seconds and a 10% load upset applied as a controller bias step (upstream of the controller output), causing a marginally stable response. After observing four response peaks, self-tune retuned the controller, stabilizing the loop. Another simulated load upset (20%) produced a well-adapted measurement response. This is followed by a 20% setpoint step with its slightly undershooting controlled measurement response.

Although the lag-to-delay ratio was halved after the pretune, causing a substantial change in the simulated-process Bode (phase) shape, the FBTUNE self-tuner worked well, reestablishing both near-optimal unmeasured-load rejection and a good setpoint response.

FEEDFORWARD CONTROL

A feedforward compensation can add to or multiply the output of an outer-loop feedback controller in order to take corrective action following a measured load upset before the controlled variable and the outer-loop feedback controller respond. The combined outer-loop feedback and feedforward output usually drive the setpoint of

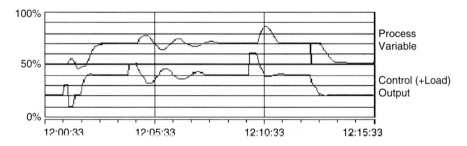

Figure 1.14   Exact MV adaptive tuning.

an inner-loop flow controller. The response of the outer-loop controlled variable to the manipulated flow is more predictable than it would be to a direct valve manipulation.

When there is no more effective delay in the manipulated variable (u) path to the controlled variable (y) than in the measured load (v) path, the measured disturbance can be rejected almost perfectly with a feedforward correction, provided the combined controller output does not limit. The ideal u/v compensator transfer function is the product of the y/v load response transfer function with u/y inverse process transfer function. Typically a gain-lead-lag-delay or simpler compensator is used to approximate the low-frequency behavior of the ideal compensator.

However, when this ideal compensator contains negative delay, the negative delay cannot be implemented and imperfect compensation results, causing the outer-loop feedback controller to react. To prevent the combined feedback and feedforward actions from overcorrecting, it is necessary to detune one. In order not to compromise the outer-loop feedback controller's ability to reject an unmeasured load upset, an additive feedforward compensator's gain should be detuned, perhaps in proportion to the ratio of the negative delay to the closed-loop characteristic time. Since the gain of a multiplicative feedforward compensator is determined by the feedback controller, it is usually desirable to avoid multiplicative compensation when the calculated compensator delay is negative.

A feedforward-type compensation can also be used to partially decouple interacting loops. The effect of the other loop's measured secondary flow can be compensated as if it were a measured external load. This approach has the advantage over matrix methods that the appropriate compensation can be applied, additively or multiplicatively, to active loops even when the other loops are saturated or under manual control.

Multiplicative feedforward compensation of a load flow is often effective for a temperature or composition outer loop manipulated with an inner-loop controlled flow. This configuration makes the process appear more linear, as seen from the outer feedback controller; thus the feedforward can be considered to provide gain scheduling for this feedback controller. Alternatively, the outer feedback controller can be viewed as adaptively tuning the gain of the feedforward compensator.

Additive feedforward compensation is often effective when the outer-loop controlled variable is level or pressure.

ADAPTATION OF FEEDFORWARD LOAD COMPENSATORS

Exact MV uses separate adaptive strategies for gain scheduling of the feedback controller and its feedforward compensators. The feedforward tuner's self-tune method determines additive and multiplicative compensator tuning for several measured loads and interacting measurements simultaneously. The moment-projection method identifies a low-frequency input/output model of the process from an isolated response to naturally occurring measured loads. No pretune is needed.

A continuous (differential equation) model, in contrast to a difference equation model, is insensitive to the computing interval h, provided h is small. A restricted-complexity model identifier, for a process that includes delay, can be based on the method of moments (Bristol and Hansen, 1987). The Laplace transform $X\{s\}$ of each signal's derivative $\dot{x}\{t\}$ can be expanded into an infinite series of moments:

$$X\{s\} = \int_0^\infty e^{-st} \dot{x}\{t\} dt = \sum_{j=0}^\infty \frac{(-s)^j}{j!} \int_0^\infty t^j \dot{x}\{t\} dt = \sum_{j=0}^\infty \frac{(-s)^j}{j!} M_{xj} \quad (1.14a)$$

Signal derivatives are used so that the moment integrals, for an isolated stable response, converge to near final values in the finite time $m \cdot h$ from the disturbance start:

$$M_{x_n^Y} = \int_0^\infty t^n \frac{dx\{t\}}{dt} dt \approx \sum_{k=1}^m \left(\left(k - \frac{1}{2}\right)h\right)^n (x\{kh\} - x\{(k-1)h\}) \quad (1.14b)$$

Higher moments are increasingly sensitive to signal noise. Therefore it is useful to low-pass filter each signal prior to the moment calculation and restrict the maximum order n.

The Laplace transform of the manipulated variable derivative U is modeled in terms of the Laplace transforms of the derivatives of the controlled variable Y and measured-load V_i with unknown transfer functions A and C_I:

$$U\{s\} = A\{s\}Y\{s\} + \sum_i C_i\{s\}V_i\{s\} \quad (1.14c)$$

Each transfer function is expanded in a series involving unknown model coefficients:

$$\begin{aligned} A\{s\} &= a_0 - a_1 s + \ldots \\ C_i\{s\} &= c_{i0} - c_{i1} s + \ldots \end{aligned} \quad (1.14d)$$

The product of the signal and model polynomials provide an equation for each power of s. These are expressed in matrix form:

Techniques for Adaptive Control

$$\begin{bmatrix} M_{i_t 0} \\ M_{i_t 1} \\ \vdots \end{bmatrix} = \begin{bmatrix} M_{\dot{y}0} & 0 & 0 \\ M_{\dot{y}1} & M_{\dot{y}0} & 0 \\ \vdots & \vdots & \ddots \end{bmatrix} \begin{bmatrix} a_0 \\ a_1 \\ \vdots \end{bmatrix} + \sum_i \begin{bmatrix} M_{\dot{v}_i 0} & 0 & 0 \\ M_{\dot{v}_i 1} & M_{\dot{v}_i 0} & 0 \\ \vdots & \vdots & \ddots \end{bmatrix} \begin{bmatrix} c_{i0} \\ c_{i1} \\ \vdots \end{bmatrix} \quad (1.14e)$$

This model structure allows the equations, moments, and unknown coefficients to be truncated. The measured moments of a given order are not affected by process coefficients of a higher order.

Each additional equation (power of s) adds more than one new unknown model coefficient. Hence there are more unknowns than equations. The projection algorithm can be used to find the smallest sum of weighted squared model coefficient changes that will satisfy a new set of signal moment observations. Expressing (1.14e) in terms of the parameter vector Θ after a new transient:

$$M_U = M_X(\Theta + \delta\Theta) \quad (1.15a)$$

where:

$$M_U^T = [M_{i_t 0} \quad M_{i_t 1} \quad \ldots]$$

$$M_X = \begin{bmatrix} M_{\dot{y}0} & 0 & 0 & M_{\dot{v}_i 0} & 0 & 0 \\ M_{\dot{y}1} & M_{\dot{y}0} & 0 & \ldots & M_{\dot{v}_i 1} & M_{\dot{v}_i 0} & 0 & \ldots \\ \vdots & \vdots & \ddots & & \vdots & \vdots & \ddots \end{bmatrix} \quad (1.15b)$$

$$\Theta^T = [a_a \quad a_1 \quad \cdots \quad \ldots c_{i0} \quad c_{i1} \quad \cdots \quad \ldots]$$

Using a Lagrange multiplier λ to enforce (1.15a), projection minimizes:

$$\Pi = \frac{1}{2}(\delta\Theta)^T(\delta\Theta) + \lambda^T(M_U - M_X(\Theta + \delta\Theta)) \quad (1.15c)$$

with the result that:

$$\delta\Theta = M_X^T(M_X M_X^T)^{-1}(M_U - M_X\Theta) \quad (1.15d)$$

Because of the leading M_X^T matrix, only those model coefficients weighting signals that are significantly active are updated.

The moment-projection approach is particularly suited for adapting feedforward gain and low-pass compensators, because the inputs need not be persistently excited. When the compensation is additive, the manipulated variable u is the sum of the feedback controller output u_{FB} and the feedforward term:

$$u = u_{FB} + \sum_i \left(\frac{c_{i0} v_i}{1 + \tau_i s + 0.5(\tau_i s)^2} \right), \quad \tau_i = \max\left\{ \frac{c_{i1}}{c_{i0}}, 0 \right\} \quad (1.16a)$$

When the compensation is multiplicative, the manipulated variable u is the product of the feedback controller output u_{FB} and the feedforward term:

$$u = u_{FB} \cdot \left(\frac{v}{1 + \tau s + 0.5(\tau s)^2} \right), \quad \tau = \max\left\{ \frac{c_1}{c_0}, 0 \right\} \quad (1.16b)$$

Only two moments need be computed for each signal and two moment equations need be solved by projection. However, when signals are cycling or responses overlap, the moment integrals do not converge in the allotted time and the adaptation must be suspended. Since an adaptive feedback controller should be capable of stabilizing a stabilizable unstable loop, the moment-projection method is not suited for self-tuning a feedback controller.

Exact MV's FFTUNE may adapt several multiplicative and feedforward compensators to reduce the effects of measured loads and interacting loops on the controlled variable. Feedback (FBTUNE) and feedforward (FFTUNE) adapters are implemented for PID and PIDτ (PIDA) controllers with Foxboro's I/A Series Exact MV and have been used to improve industrial process performance since 1994. Both adapters adapt gain schedules. Based on conditions when a new disturbance is detected, the most appropriate stored set of tuning constants is inserted into the feedback and feedforward controllers and updated if necessary when enough of the response has been analyzed.

Figure 1.15 illustrates the workings of an Exact MV PIDA controller and FBTUNE and FFTUNE tuners on a simulated process using an I/A Series control processor. The lower curve is the controller output. The upper curve is the controlled variable. First the feedback tuner's Pretune applied a doublet pulse to the open-loop process in order to identify the process and achieve initial feedback-controller tunings.

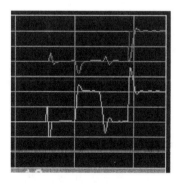

Figure 1.15 Exact MV feedback and feedforward tuner results.

Then the loop was closed and a measured-load step was applied. This triggered the start of moment calculations by the feedforward tuner. Because the feedforward compensator started with zero parameters, the controlled variable responded to feedback as if the load were unmeasured. The feedback tuner's self-tune did not update the feedback-controller tunings because the desired response shape was achieved. When the signals settled, projection updated the feedforward compensator coefficients.

Next an equal and opposite measured-load step was applied. The adapted feedforward compensator substantially reduced the controlled variable's response. Usually the good compensator settings are achieved after the first or second adaptation. However, more adaptations may be required if several measured loads change simultaneously. Finally, a setpoint step was applied.

CONCLUSION

The stability of a loop with no steady-state error and an uncertain delay cannot be guaranteed solely by limiting its $\|h\|_\infty$ norm or by specifying the smallest distance of the nominal Nyquist plot from the -1 point (e.g., Figure 1.2). Small delay changes can cause large high-frequency phase shifts. Process gain and frequency (time scale) margins are more reliable robustness measures.

A controller designed for unmeasured load rejection should generally apply larger proportional and derivative actions to its measurement than to its setpoint. Derivative should not be applied to setpoint and should not be used at all for a pure delay (or an all-pass) process.

Tuning of a PID controller for good load rejection should not be based on the loop's setpoint response. Good setpoint response, but not good unmeasured load rejection, can be achieved without feedback. Manual tuning is more difficult when the process has a dominant lag than a dominant delay because adjustment of its gains in either direction may make the loop more oscillatory.

Loop delay limits the ability of a controller to reject unmeasured load. Optimal integral and derivative times are proportional to the delay. The minimum IAE is proportional to the effective delay raised to a power equal to one more than the number of dominant lags. Identification of effective delay in the presence of dominant lags requires process excitation that emphasizes relatively high frequency behavior (near potential loop unity-gain crossings). An open-loop step or a closed-loop nonoscillatory response may not be good enough.

An application-proven robust adaptive method for tuning a PID controller gain schedule is presented. It uses an open-loop doublet pulse response method to initially identify a process model and a simple and direct algebraic tuning method to pretune the controller to reject an unmeasured load. In closed-loop operation the gain and time scales of the currently active model and controller are self-tuned based on observations of natural loop oscillations. Gain scheduling inserts the most appropriate stored tuning set when a new disturbance is detected.

A simple and direct algebraic method of calculating PID tuning constants for unmeasured load rejection based on a linear differential-equation process model (including time delay) is presented. Methods are discussed for experimentally identifying the process model from an open-loop doublet-pulse response and for robustly updating the process model from a closed-loop oscillatory response to an unmeasured load disturbance. These methods are combined to provide field-proven self-adaptive PID gain scheduling.

A PID or deadtime (PIDτ) controller, designed for maximum unmeasured-load rejection and nonovershooting setpoint response, applies less (or equal) gain to setpoint than to measurement and no derivative to setpoint. An algebraic tuning method is presented for tuning these controllers. The IAE of the PIDτ is significantly less than that of the PID, but the PIDτ is less robust.

A pole-cancellation or model-feedback controller is capable of faster response to setpoint provided its output does not limit, but both are poor at rejecting unmeasured load when the process has a dominant lag. These controllers apply all their terms to the control error. Feedforward compensation can significantly reduce the controlled variable's response to measured loads or interacting control variables. Adaptive tuning of compensator gain schedules from responses to naturally occurring disturbances enables the compensators to retain their effectiveness in spite of process nonlinearity and time shifting behavior.

REFERENCES

Bristol, E. H. (1966). "On a New Measure of Interaction for Multivariable Process Control," *IEEE Trans. Autom. Control*, January, pp. 104–106.

Bristol, E. H., and P. D. Hansen (1987). "Moment Projection Feedforward Control Adaptation," Proc. 1987 American Control Conference, p. 1755.

Hansen, P. D. (1998). "Controller Structure and Tuning for Unmeasured-Load Rejection," Proc. 1998 American Control Conference, Vol. 1, pp. 131–136.

Panagopoulos, H., K. J. Åström, and T. Hagglund (1998). "Design of PID Controllers Based on Constrained Optimization." Department of Automatic Control, Lund Institute of Technology, Box118, S-221 00 Lund, Sweden.

Shinskey, F. G. (1994). *Feedback Controllers for the Process Industries*. McGraw-Hill, New York.

SUGGESTED READING

Åström, K. J., and B. Wittenmark (1984). *Computer Controlled Systems*. Prentice-Hall, Englewood Cliffs, NJ.

Hansen, P. D. (2000). "Robust Adaptive PID Tuning for Unmeasured-Load Rejection," Proc. 2000 IFAC Workshop on Digital Control, pp. 487–494.

Hansen, P. D. (1997). "Process Control" section in *Electronics Engineer's Handbook*, 4th ed. (D. E. Christiansen, Ed.). McGraw-Hill, New York.

2

THE EXPLOITATION OF ADAPTIVE MODELING IN THE MODEL PREDICTIVE CONTROL ENVIRONMENT OF CONNOISSEUR

David J. Sandoz

Connoisseur is a software product for the design and application of industrial model predictive control (MPC). It is owned and marketed by the Foxboro Company, part of the Invensys group of companies. Connoisseur has been employed in industry since 1984 (Sandoz, 1984), and in those early days, one of the stronger features in its range of offerings to the marketplace was the capability to implement adaptive modeling in support of multivariable predictive controllers. This technology has been the backbone of Connoisseur for many years.

Connoisseur has evolved a great deal since 1984, and it is now Foxboro's main platform for advanced control (Sandoz et al., 1999). The capability of the control engineering of Connoisseur has progressed at a great pace, in line with the general advances in MPC technology that have taken place in the past two decades. The character of the adaptive modeling within Connoisseur has also changed, with many features being added to enhance flexibility and applicability. It forms one aspect of a comprehensive tool set for the structuring and statistical identification of process models that describe the cause-and-effect dynamic behavior of industrial plant. However, at its core, the basic principles that are exploited today are little changed from those of 1984 and earlier (Hastings and Sage, 1969; Sandoz and Swanick, 1972). The adaptive modeling of Connoisseur employs recursive least squares (RLS) mechanisms to characterize and update the coefficients of multivariable sampled data models in real time. This real-time modeling approach has always been a very important capability within Connoisseur and has always been recognized as a key factor in the success and sustainability of the Connoisseur product. It is rare for the adaptive modeling facility not to be used in some context when Connoisseur is applied to control an industrial process. Connoisseur encapsulates a great deal of experience in the use of such technology.

Given these facts, it remains a puzzle to the author that the exploitation and use of such adaptive methods has not been more widespread. Even today, despite the well-known capabilities of Connoisseur in real-time modeling, many of the major competitors to Connoisseur only offer offline modeling facilities (Qin and Badgewell, 1997).

The adaptive modeling of Connoisseur was established toward the end of an era that saw much academic input into the area of adaptive control. This was the period during which well-known figures such as Åström and Clarke were developing and evaluating self-tuning control concepts (Åström and Wittenmark, 1989; Clarke and Gawthrop, 1975). It was the period of the early days of microprocessing and a number of products were brought to market to offer a combination of the bright new hardware and software technologies to industry. Unfortunately not many of these have proved sustainable. One particular product that has held ground is "Novatune"—a self-tuning controller developed by ABB in association with Åström (Åström and Wittenmark, 1989). Novatune algorithms are still offered today as part of the range of controllers that can be called up within ABB DCS computers. The algorithms used within Connoisseur have much in common with those of Novatune.

The adaptive modeling of Connoisseur was first tested in 1984 in application to a gasoline engine mounted on an engine test bed (Hesketh and Sandoz, 1987). The objective was quite simple: to simultaneously control the torque and speed of the engine, using brake and throttle as manipulated variables (MVs). The model structure was ARX in character (see below) with a prediction interval of 0.5 sec and just 12 coefficients in the model. The software ran on a PDP11 computer and all coding was in Fortran. The project was a great success. It took just 15 minutes to tune in a model, with two simultaneous PRBS sequences being applied to the MVs. The customer was delighted with the performance of the associated controller, which was capable of driving the engine in ways that had never been possible to achieve with conventional PID controllers. Today, adaptive modeling continues to provide a powerful service to the customers of Connoisseur, although in a much more comprehensive framework of modeling and control.

The adaptive capabilities of Connoisseur modeling have of course progressed considerably since that time, for example, to encompass nonlinear systems. A particularly useful and powerful approach involves radial basis functions (RBFs) (Haykin, 1994).

The most important aspect of RBFs in the context of this paper is that the linear RLS identification procedure can be employed to directly characterize the coefficients of an RBF model. The benefit of this in providing a consistency of style to a process engineer, whether modeling in the linear or the nonlinear domain, is very important. Of course, this importance is seriously compromised if the RBF approach is deficient in comparison with other neural network formats. Comparative studies with, for example, multilayer perceptron structures (which require back propagation/hill climbing methods for the identification of coefficients) suggest that RBF models are at least of equal capability in describing nonlinear process dynamics. In fact these studies have culminated in the somewhat pedestrian conclusion that the simplest form of RBF structures that involve the allocation of random centers without any complexity of internal feedback or clustering, are the most efficient and reliable for use with model predictive control. Another modeling approach that is now developing with Connoisseur is that of partial least squares (PLS) (Geladi, 1988). This method is more usually employed to form a basis for the use of multivariate statistics for condition monitoring. However, certain attractive properties that are features of PLS make for its appropriate use, in certain circumstances, in the control engineering environment.

The first section of this chapter first of all reviews the general character of models of linear process dynamics that are used for the design of control systems. It is then concerned with the generalized model structure that is employed within Connoisseur for control system design. Consideration is given to the various ways in which this structure can be configured and their relative merits for modeling and for control.

The first of the next two sections reviews the selections that an engineer has to make in defining a model structure. The second is concerned with methods for the identification of model coefficients and includes a review of the recursive least squares approach around which adaptive modeling pivots.

The subsequent section is concerned with the nuances of adaptive modeling and with the practical facilities that have been introduced into Connoisseur to enable an engineer to best appreciate online performance and to interact appropriately to manage the online operations.

The next section is concerned with the nonlinear radial basis function (RBF) modeling facility of Connoisseur and with the PLS facility.

The final sections begin with a short description of a model of a fluid catalytic cracking unit (FCCU) followed by a demonstration of the capabilities of real-time adaptive modeling in application to this simulation. A small model predictive controller is installed to regulate some critical parameters of the FCCU. Adaptive modeling is demonstrated in the exercise of updating and refining the model that is being used by the predictive controller. The subsequent section describes how an initial model is obtained, by carrying out preliminary tests on the process, in order to establish a working controller in the first place. Next it is shown how individual cause-to-effect paths may be readily modeled in real time, to extend the scale of the working controller to be able to manage other process variables. The last section describes an exercise to completely regenerate a model for use with the controller, during the course of the operation of that controller, and which illustrates the exploitation of multivariable adaptive control.

MODEL STRUCTURES

Model Forms for Representation of Linear Process Dynamics

Industrial Process Dynamics may be characterized by a variety of different model types (Kuo, 1982). Signals for such models are usually grouped into three categories:

- Controlled variables (CVs): plant outputs or effects that can be monitored
- Manipulated variables (MVs): plant inputs or causes that can be monitored and manipulated
- Feedforward variables (FVs): plant inputs or causes that disturb the process but that can only be monitored

The most simple model form for relating the CV response y of a process to a change in the MV input u is

$$y(s) = G(s)u(s) \qquad (2.1)$$

where s is the Laplace operator and $G(s)$ is the transfer function. For a second-order system represented parametrically, a general form is given by

$$G(s) = G(1 + Tn)e^{-sTd}/\{(1 + sT_1)(1 + sT_2)\} \qquad (2.2)$$

with G the process gain, Td the pure time delay and T_1 and T_2 the time constants. Tn, the numerator time constant, permits the description of nonminimum-phase (inverse response) behavior. If the response is underdamped, the denominator will take the form

$$\{1 + 2\xi w_n s + (w_n s)^2\}$$

with ξ the damping ratio and w_n the natural frequency. If the response is integrating, then an additional s will multiply into the denominator.

It is the case for most industrial processes that Equation (2.2) provides an adequate basis for describing the cause-to-effect behavior of the majority of single-input-to-single-output paths taken in isolation (note that if the dynamics for a path are simple and first-order, then T_2, Tn, and Td will be 0). A complex process with many inputs and outputs could involve hundreds of such transfer functions. Thus the more general equation prevails as

$$\mathbf{y}(s) = \mathbf{G}(s)\mathbf{u}(s) \tag{2.3}$$

with $\mathbf{y}(s)$ and $\mathbf{u}(s)$ vectors composed of multiple CVs and MVs (and FVs), respectively, and $\mathbf{G}(s)$ a matrix of transfer functions.

The time domain equivalent to Equation (2.2) may be given by two equations, the standard state-space differential equation and an associated measurement equation. A simplified form of these is

$$d\mathbf{x}(t)/dt = \mathbf{A}\mathbf{x}(t) + \mathbf{B}\mathbf{u}(t) \tag{2.4a}$$

and

$$\mathbf{y}(t) = \mathbf{C}\mathbf{x}(t) \tag{2.4b}$$

with $\mathbf{x}(t)$ the state vector of dimension n (order of dynamics), $\mathbf{A}$ the transition matrix, $\mathbf{B}$ the driving matrix, and $\mathbf{C}$ the measurement matrix.

Variations of Equations (2.3) and (2.4) commonly feature in the control engineering literature as the basis for developing control algorithms.

Linear Models for Control Engineering Design

Industrial control engineering generally exploits models that are in sampled data form and consideration normally restricts to models of two basic categories: parametric state-space (commonly termed ARX—short for autoregressive–exogenous) or nonparametric finite-step/finite-impulse response (commonly termed FSR/FIR). In fact both of these categories may be subsumed by one general-purpose multivariable time-series format subject to the presence and scale of various dimensions. It is possible to derive such an equation directly from either Equation (2.3) or Equations (2.4) above (Sandoz and Wong, 1979). The equation that arises has the general form

$$\mathbf{y}_{k+1} = \mathbf{A1Y}_k + \mathbf{B1U}_k + \mathbf{C1V}_k + \mathbf{d} \tag{2.5}$$

with

- $\mathbf{k}$ the sampling instant, and $\mathbf{k} \to \mathbf{k}+1$ the sampling interval T
- $\mathbf{y}$ a vector of p CVs that are predicted by the model
- $\mathbf{Y}$ a vector of R samples of $\mathbf{y}$, that is,

$$Y_k = |y_k^T, y_{k-1}^T, \ldots, y_{k-R+1}^T|^T,$$

with the dimension of **Y** being $\geq n$, the order of dynamics

- **U** a vector of S samples of **u**, that is

$$U_k = |u_k^T, u_{k-1}^T, \ldots, u_{k-S+1}^T|^T,$$

with **u** a vector of m MVs

- **V** a similar vector of S samples of **v**, with **v** a vector of q FVs.

The dimensions of the transition matrix **A1** and the sampled vector **Y** are set by the number of CVs and the orders of process dynamics that prevail for each path of the model (i.e., single-input cause to single-output effect). The transition matrix may be considered as a composite of all of the transfer function denominators associated with these paths. The offset vector **d** is not required if differences in the samples of **y**, **u**, and **v** are taken across each sampling interval (i.e., using $y_k - y_{k+1}$ rather than y_k, etc.). A model is termed incremental if differences are taken and absolute if not.

The discrimination between whether to choose an incremental or an absolute format can depend on the character of the data that is to be used for identification, the character of the process that is to be identified, or the purposes to which the final model is to be put. For example, the differencing procedure amplifies the noise-to-signal ratio. This makes the identification less accurate in particular in characterizing the effects of long-term transients if the data is noisy, if the model structure is inappropriate, or if the process is nonlinear. On the other hand, if the process is nonstationary or nonlinear, the absolute approach can be grossly misleading. The choice made has to be taken on a case-by-case basis.

The transition matrix **A1** can be presented in two forms. The first, termed homogeneous, restricts the prediction relationship for any CV to include only sampled history for that particular CV and not for any other. The second, termed heterogeneous, bases the prediction of each CV on reference to the sampled history of all of the other CVs. The heterogeneous form is consistent with the standard state space equation (Equation 2.4). The usual form adopted with chemical plants is homogeneous, because CVs are often distributed quite widely around the process and dynamic interdependence among the CVs does not necessarily make practical sense. Another advantage to the homogeneous form is that if a CV fails, it does not affect the ability of the model to predict behavior in the other CVs. With the heterogeneous form, the failure of any CV will invalidate the ability to predict for every CV of the model.

The dimensions of the driving matrices **B1** and **C1** are set by the number of MVs and FVs and by however many samples S are necessary in **U** and **V** to establish a model that accurately reflects the process. This vague statement firms up very easily if the equation is representative of an FIR or FSR structure. The transition matrix is removed (i.e., the order of dynamics may be considered to be zero on all paths so that dynamic behavior is represented entirely by the transfer function numerators). The number of samples for any path is then as many as required to encompass the time span to settle to steady state following a step change in any input (i.e., the sum total of any pure time delay and dynamic transients). The reason for the term "finite" is that the response is truncated at some stage, usually at the point where the contribution from extra terms is insignificant. The overall dimensions of **B1** and **C1** are thereby defined by consideration of the paths with the maximum sampling requirements and the matrices are padded with zeros to maintain validity. The value of S is, in effect, different for each cause signal and the overall matrix formulation of Equation (2.5) is maintained by such padding. It can be that the matrices are quite sparse. A mechanism of cross-referencing pointers is then desirable and this can dramatically reduce workspace storage demands (which can be very large, particularly when the model involves a mix of fast and slow responses).

If a transition matrix is declared, then in principle the dimensions of the matrices **B1** and **C1** should be significantly reduced. These matrices are then no longer representative of system time responses but correspond to the driving terms of a set of difference or differential equations, such as in Equation (2.4). In this case, for any input MV or FV, there must be at least as many samples as are needed to span the largest time delay into any of the CVs; otherwise the equation would not be causal. A few extra samples may be necessary in order to validate the description of dynamics, should this be complex. Further, if there are multiple delayed impacts from an input, for example such as arise with processes that involve the recycling of material streams, then the sampling must extend to cover all of the successive impacts.

Many industrial control engineering technologies restrict to the use of the FIR/FSR form (Qin and Badgewell, 1997), with the implication of very large model structures, sometimes for situations that are quite simple.

The reasons for this preference are twofold:

- An engineer is able inspect the pattern of the FIR coefficients to gain a feel for the time constants and gains of the process.
- There is no need to be concerned with the selection of the "order" of a transition matrix and the basis for selection of the number of terms in the driving matrices is clear, as indicated above.

Large structures impose significant computational burden in solving for control moves, but this is of little consequence except for very large systems, given the state of today's low-cost and high-performance computer power. However, such structures do create problems for statistical identification methods because of the large number of coefficients that have to be determined. The identification of many coefficients requires large amounts of data. This imposes a time constraint on delivering good results, irrespective of computer power (e.g., by requiring extended exposure to plant for data collection and experimentation).

The state-space parametric or ARX forms can be much more compact and therefore more suited to efficient identification of their parameters. However, there are disadvantages. Accurate prediction for such a model is dependent on good reflection of dynamics within the sampled history of the CVs. If these signals have significant levels of noise superimposed on them, then, subject to sampling intervals and the capability to employ suitable filtering, the ability of the model to accurately predict steady state from a referenced dynamic state can be compromised. In addition, multivariable structures that involve both fast and slow dynamics can present difficulties. For example, it is very common for a controller to involve CVs that are liquid levels, which respond in seconds, and other CVs that are analyzers monitoring some chemical composition, which respond in tens of minutes. Proper description of the dynamic behavior of the levels requires a very short sampling interval. This short interval must then be imposed on the model segment that describes the analyzers. Accurate representation of analyzer behavior then relies on a high degree of precision in the transition matrix coefficients. Such high accuracy is not practicable, particularly when the process signals are noisy or nonlinear.

The form of model used also has practical implications for the use of the model for prediction, particularly with respect to initializing a model in the first place. The very first prediction that a model makes can only be correct if the sampled history in the model vectors properly reflects earlier plant behavior. If the vectors are large, waiting time before a model can go online can be significant, with as much as 1 or 2 hours being quite common. Of course, this aspect is irrelevant if the model prediction is initiated with the process in a steady-state condition, in which case all of the earlier samples can be set to the current values.

The proper description of processes that involve CVs that have integrating properties (e.g., as is often the case with liquid levels) or that are open-loop unstable necessitates the involvement of a transition matrix **A1**. Coefficients within **A1** then reflect properties that are equivalent to poles being on or outside the unit circle (in z domain parlance).

ISSUES FOR IDENTIFICATION

Selection of Best Model Structure

The questions that an engineer is faced with in selecting a model structure for identification may be summarized as follows:

- What cause and effect signals should be incorporated?
- What prediction interval T is appropriate?

- Should the model be of ARX or FIR structure?
- If ARX structure is selected, what order of dynamics should the transition matrix have (i.e., value of R), and should the model be homogeneous or heterogeneous?
- How many samples of MVs and FVs are needed (i.e., value of S)?
- Should the model be of incremental or absolute structure?

These choices have to be made in conjunction with any Identification procedure that is employed to determine the coefficients of the matrices **A1**, **B1**, **C1**, and **d** of Equation (2.5). Proper selections can be critical to being able to establish an effective cause-to-effect description of process behavior.

An engineer facing an industrial process with the task of developing a multivariable predictive model will normally, first of all, choose the FIR approach. The interval T will be set to some factor of the shortest time constant (say one-tenth) and the number of samples S will be set to span three times the largest time constant. The cause and effect signals will be selected on the basis of process knowledge and other design objectives (e.g., the requirements of a control system). Such specifications become second nature and intuitive to an experienced application engineer. The engineer will design an experiment to generate data that can be used to identify appropriate coefficients for Equation (2.5). This may involve the application of a series of step tests, with each MV being taken in turn, or it may involve the simultaneous manipulation of a number of MVs by employing special white-noise sequences, such as PRBS (pseudo-random binary sequences). The data generated is then analyzed, with the engineer steering the analysis interactively to establish the best model. Such interaction will involve successive model identifications with variations being made systematically in the model structure subject to the judgment of the engineer—that is, the identification determines the coefficients and the engineer determines the structure.

These selections, although they simplify decision making for the application engineer, are not necessarily the most appropriate for statistical identification procedures that are used to analyze plant data to determine the matrix coefficients. This is particularly the case if preference is to use real-time adaptive modeling to characterize model parameters in parallel with the real-time operations of control systems. This issue is discussed in more detail in the next section. ARX formulations may be more appropriate because much less data might be needed for identification and because the compact format might give rise to more effective control systems. However, the determination of structure is not intuitive, as is the case for the FIR formulation. One approach that has been provided in the Connoisseur product may be employed after an FIR model has been approximately determined. The facility is available to transform between the FIR and transfer function representations (Equations 2.5 and 2.2). Each cause-to-effect path in Equation (2.5) is taken in turn. The best second-order transfer function (see Equation 2.2) that approximates the FIR relationship is determined. This is achieved by using a least-squares approach that estimates gain, time delay, numerator time constant, and denominator time constants (or damping ratio and natural frequency if underdamped). These individual transfer functions provide the basis for selection of S for each cause signal and for selection of the process order n, in Equation (2.5). The coefficients for an ARX representation of Equation (2.5) may then be calculated from the original FIR terms, again employing a least-squares approach to give best approximation. The loss of accuracy in this two-stage transformation is dependent on the ability of the second-order transfer function to accurately describe process behavior. In practice and fortunately in most process plants, most cause-to-effect paths can be effectively represented by the generality of Equation (2.2).

Identification Methods

The issue of identification of predictive models is, given a declared structure for Equation (2.5), to determine the most accurate representation for the coefficients of the associated matrices, usually by direct application of statistical methods to recorded process data. Special tests are normally applied to the process to generate such data, as mentioned above. Before any exercises toward such identification commence, it is of vital importance that the process be set in the condition that will prevail for normal and subsequent operations. Thus the base layer of conventional regulatory control systems should be properly tuned and the controllers should be in automatic mode. Any models obtained will encapsulate these control systems, and if a controller is switched off or even retuned, predictive models of the process might become invalid. It is also important that any high-frequency noise associated with CVs and FVs be eliminated as close to source as possible (e.g., by low-pass filtering in the data-gathering computers) prior to being sampled for identification. Improper filtering can give rise to inappropriate noise with aliasing and can complicate the identification process at a later stage.

The subject of identification is a broad one and there are many methods and approaches (Ljung, 1987). However, there are a small number of techniques that are most often exploited in association with control engineering. These techniques are based on the least-squares principle and involve a degree of processing to reduce bias arising because of measurement noise associated with the CVs. In fact, in identifying coefficients for an FIR/FSR model directly, problems of bias because of noise are of less consequence because of the lack of a transition matrix, which is to some extent a counterbalance to the problems arising because of the excessive number of parameters.

Incorporating P samples of data, Equation (2.5) may be rewritten in the form

$$\mathbf{w}_k = \alpha \mathbf{Z}_k \qquad (2.6)$$

with

$$\mathbf{w}_k = [\mathbf{y}_{k+P}, \mathbf{y}_{k+P-1}, \ldots, \mathbf{y}_{k+1}] \text{ and } \mathbf{Z}_k = [\mathbf{z}_{k+P-1}, \mathbf{z}_{k+P-2}, \ldots, \mathbf{z}_k]$$

where $\mathbf{z}_k$ is the aggregation of vectors $\mathbf{Y}_k$, $\mathbf{U}_k$, and $\mathbf{V}_k$, and α is that of the matrices **A1, B1, C1**, and **d**.

The ordinary least squares (OLS) estimate of α is given as

$$\alpha_k^{LS} = \mathbf{P}_k \mathbf{Z}_k^T \mathbf{w}_k, \text{ with } \mathbf{P}_k = [\mathbf{Z}_k \mathbf{Z}_k^T]^{-1} \tag{2.7}$$

There are numerical procedures that can solve for α_k^{LS} in a very efficient manner; however, OLS has its limitations. First, if the matrix $\mathbf{Z}_k \mathbf{Z}_k^T$ is singular, which will arise if there is correlation between the parameters in $\mathbf{Z}$, then a solution cannot be obtained. Such a situation is not uncommon. Second, if the model is of ARX structure, OLS produces biased answers if the CVs, that is, the terms of the AR vector $\mathbf{Y}$ of Equation (2.5), are subject to measurement noise. Thus OLS would normally be used for FIR structures and with care to be sure that correlated data is avoided.

There is another approach that does not suffer from these deficiencies but that involves more computation—that of recursive least squares (RLS). In the context of industrial exploitation, one RLS algorithm in particular stands head and shoulders above all others—Bierman's UD Filter—UD meaning upper diagonal (Bierman, 1976). This algorithm, in association with some simple predictive filtering, provides a very powerful and robust tool for model identification. The filtering mechanism, essentially an instrumental variables approach, is able to cut through bias that is associated with noise on the CVs. The mechanism for RLS is indicated by Equations (2.8):

$$\alpha_{k+1}^{LS} = \alpha_k^{LS} + (\mathbf{y}_{k+1} - \alpha_k^{LS} \mathbf{z}_k^f) \mathbf{P}_k / d_k$$

$$\mathbf{P}_{k+1} = \mathbf{P}_k - \mathbf{P}_k \mathbf{z}_k^{fT} \mathbf{z}_k^f \mathbf{P}_k / d_k$$

$$d_{k+1} = \rho + \mathbf{z}_k^{fT} \mathbf{P}_k \mathbf{z}_k^f \tag{2.8}$$

$$\mathbf{y}_{k+1}^f = \alpha_k^{LS} \mathbf{z}_k^f$$

Equation (2.8) computes a new least squares estimate α_{k+1}^{LS} on the basis of the previous estimate α_k^{LS}, the covariance $\mathbf{P}_k$, and new plant information in $\mathbf{y}_{k+1}$ and $\mathbf{z}_k^f$. The elements corresponding to $\mathbf{y}_k$ in $\mathbf{z}_k^f$ are updated with the filtered values $\mathbf{y}_k^f$ that were computed as $\mathbf{y}_{k+1}^f$ at the previous iteration. To start the algorithm, with $k = 0$, then $\mathbf{P}_k$ is set to a diagonal matrix, with very large values on the diagonal, and α_k^{LS} is set at zero. The UD filter algorithm is a sophistication of Equations (2.8) that incorporates upper diagonalization of the symmetric matrices so that it is not possible for the matrix **P** to ever arise as negative definite—making the recursive computations very robust.

The parameter ρ is the forgetting factor, which is always set to 1 in Connoisseur. The forgetting factor concept is of relevance when the RLS algorithm is employed in real time for adaptive modeling. The idea of a forgetting factor is to allow the RLS algorithm to slowly forget older data, placing maximum emphasis on recent data. To achieve this, it is suggested in the literature that the forgetting factor be set to a value just less than 1 (e.g., 0.99). This is a bad idea! If the plant is not moving, there is no new information coming in and the healthy model parameters computed from old data will be slowly erased. There is, however, a need to have a facility to encourage model updating. In Connoisseur, this is achieved by imposing values onto the diagonal of the covariance matrix **P**, termed "jolting the adaptor." This is done as the result of a deliberate command from the engineer, rather than automatically. Increasing the covariance values is tantamount to decreasing the confidence in the model parameter estimates. This has similar influence to a forgetting factor except that the influence quickly decays away so that model estimates stabilize even if excitation is only small.

Another major advantage of the RLS approach is its ability to deliver effective results despite the presence of correlated data within the data sets to be analyzed. This is a feature that has long been trumpeted as a powerful feature of partial least squares (PLS), which is considered later, but has not been widely realized to be at attribute of RLS. Thus, for example, if two FVs are highly correlated, the covariance matrix that is calculated as part of the least squares procedure (i.e., $\mathbf{Z}_k\mathbf{Z}_k^T$ in Equation 2.7) is likely not to have full rank (i.e., to be nonsingular). OLS is required to invert this matrix and will fail unless it has full rank. RLS, in contrast, does not invoke such inversion. Although the results from RLS will not properly describe the individual contributions from the correlated signals, but rather some aggregated effect, the contributions of other signals will be properly described.

RLS is probably the backbone of most self-tuning controllers and has been traditionally recognized as a powerful engine for adaptive modeling. This is because the identified coefficients are updated sequentially with each progressing set of data values and because the filtering mechanism is very stable—an ideal procedure for real-time implementation. However, the above factors make RLS a powerful engine for least squares modeling in general, not just for model adaptation in real time. It is certainly exploited to full effect in this sense within Connoisseur and is probably used widely in other industrial products.

It may not be possible to directly manipulate all inputs to a process. Some disturbing influences will only be measurable (i.e., not able to be manipulated) and therefore cannot be presented as steps or PRBSs. The determination of a suitable model for such FVs can be a real problem. Timing of such a disturbance might be arbitrary. It might be of such magnitude as to necessitate adjustment of other MVs in order to avoid operational difficulties (this will unfortunately then disguise the proper impact of the disturbance because of the correcting influence, i.e., an unfortunate consequence of feedback). An RLS procedure employed in real time, updating model coefficients in sympathy with incoming plant data, provides a very effective basis for dealing with this issue. The procedures can be set up to catch and process such disturbances as and when they arise during normal process operations and to compensate for the influence of feedback. These aspects are dealt with thoroughly in a later section.

No matter how good a statistical identification procedure is, it will not deliver effective results unless presented with rich process data that reflects the issues of dynamics and steady state that are desired to be modeled. Some control engineering technologies require that a model reflect both the short-term dynamics and the steady state with a good degree of accuracy (all technologies are advantaged if this is the case) because the issues of dynamic control and steady-state optimization are entwined within their particular procedures. In such a case, there is little alternative other than to apply careful step testing to the process so that any long transients into steady state are properly picked up (such long transients are particularly prevalent with distillation processing that incorporates large-scale heat integration). The problem with such step testing is that it is required that the plant be operated open-loop, with many of the important control loops switched off, and that disturbances from the environment be kept small, if this is possible. As a general rule a number of tests are applied individually to each cause-to-effect path in the multivariable process. Such testing is expensive and can fully consume the skills of the process control and operations staff. It is a necessary evil for all predictive modeling and, if possible, is best kept to a minimum.

Pulse testing, rather than step testing, is often more appropriate if the process involves integrating paths, allowing rich data to be generated without the process going out of bounds. From the perspective of the model, issues of integration can be avoided by incorporating rates of change of CVs rather than the absolute values, although this may not necessarily be consistent with control engineering requirements. From the perspective of control the inclusion of both a CV and its rate of change can be beneficial for the management of aspects with very slow rates of integration, such as often prevail, for example, with temperatures in heating systems. The problem with such systems is to obtain representative test data for quality identification.

An alternative or supplement to step testing is the use of white-noise excitation to generate rich data for process modeling. In practice this usually means the use of pseudo-random-binary-sequence (PRBS) excitation. A PRBS can be implemented to "tickle" a process over and above any manipulations made by the operator so that a semblance of normal production can be retained during the testing process. A number of distinct PRBSs can be applied simultaneously to a number of process MVs as long as the various sequences employed are orthogonal to each other in the statistical sense. The definitive white-noise signature of each PRBS is a distinct advantage to the statistics in discriminating between deliberate and unknown disturbing influences. Quality identification is possible despite the impact of unmeasured disturbances, which is not the case if step tests are applied. Further, PRBS excited data is most appropriate for effective analysis using other statistical methods, such as correlation and Fourier methods. The disadvantage of PRBS application is that it is unlikely to accurately reflect any steady state because of the persistent excitation. To better reflect longer time transients in PRBS-generated data, it is often expedient to vary the mean of the generated sequences from time to time during the course of their application.

The identification of model coefficients goes hand in hand with the determination of model structure that was reviewed in a previous section. It is essential that effective interactive procedures be available to allow the engineer to systematically apply judgment in assessing the effectiveness of various selections and the consequential implications on the accuracy of identified models. Such interaction requires attractive graphical user interface facilities that provide a variety of services. For example, some of the services available in Connoisseur are:

- Visualization; of trends of plant data together with model prediction and error signals; and of model coefficients, with capabilities to alter and cut and paste.
- Data-editing; filters, removal of spikes and outliers, application of non-linear algebraic expressions, removal of inappropriate periods of data.
- Aids to assist the tuning of the base layer of PID control systems. A gain and phase-margin based frequency response design facility is available for determination of the most appropriate settings for PID terms. This can be presented with gain, time constant, and time delay information in order to produce recommendations, or interpretation can be made directly for any cause-to-effect path of the model form of Equation (2.5).
- Auto and cross correlation analysis for determination of impulse responses, sampling rates, time delays and parameter associations.
- Power spectrum analysis for determination of sampling rates and detection of aliasing.
- Provision for the generation and management of PRBS applications, both for initial open-loop testing and for the real-time support of adaptive modeling.
- The capability to translate between state-space/FIR/FSR model structures and simplified transfer function representations $G(s)$ in the s domain, as mentioned previously. This provides the engineer with a well-understood

basic representation in terms of gains, time constants, time delays, etc., against which process judgment and common sense may be exercised in assessing the credibility of a model.
- Residual analysis tools that provide for the contributions of a model to be assessed and the character of modeling errors to be analyzed, including the ability to assess model performance against data that has not been used for the identification calculations.
- The capability to mask off and preserve segments of a model during the identification process so that a model can be enhanced as appropriate when new data becomes available. Segments of a model that are valid can be preserved while other segments are replaced. This facility takes on extra significance when considered for adaptive modeling.
- The capability to mask off segments of data within a data set that is being processed for identification. There may be periods within a data set that are not appropriate for use, for example corresponding to a time when the process was not operating properly.

ADAPTIVE MODELING

Adaptive modeling is concerned with the updating of model coefficients, in real time, so that if process characteristics alter, the model keeps track of these alterations.

There are three primary motivations for using adaptive modeling in a model predictive control (MPC) environment.

- MPC can only perform with quality if the model employed for control system design is a reasonably accurate representation of the process that is being controlled. A very common experience is that good models are obtained in the first instance on the basis of the initial testing and data collection. However, in time model accuracy deteriorates because of changes to the process, changes in operating conditions, general wear and tear, etc. Control performance therefore deteriorates. In fact, controllers are often deliberately detuned so they can keep working despite the deterioration. At a certain point, a decision has to be made to repeat the modeling tests on the plant so that the effectiveness of the MPC may be restored, or to simply stop using the controller. Unfortunately, all too often, it is the latter choice that is made. Adaptive modeling, which can be applied in parallel with the closed-loop operations of MPC, provides the potential for maintaining the quality of the models used by the controller without the need for repetition of special, time-consuming, and expensive open-loop tests.
- Adaptive modeling can also be used to build new cause-to-effect relationships that are additional to those currently being used for control. This can be done without need to stop the controller and, in many cases, without the

need to apply special test signals. This aspect is particularly useful in modeling the influence of FVs that could not be incorporated into the early program of open-loop testing. A distinct and powerful feature of the technology of Connoisseur is the ability to characterize such disturbances while MPC is in progress, using identification in real time and using control models to filter out the effects of feedback prior to the analysis of the disturbing path.

- The RLS identification method (see Equations 2.8) is an appropriate procedure for adaptive modeling since it progresses sample by sample, with model estimates being updated at each sampling instant on the basis of the updating samples of cause and effect signals from the process. If the model predictive control calculations always use the latest model estimates as they become available, then the combination of identification and control may be classified as an adaptive control system or a self tuning control system. However, the character of the control management is much more complex than is treated in classical self-tuning control considerations.

Some major factors in considering adaptive modeling in conjunction with MPC are:

- *The influence that control feedback action has on the modeling process.* MPC is required to use the open-loop model that relates cause-to-effect. However, the control action itself imposes an additional transfer function from effect to cause. In other words the dynamics of closed-loop behavior are different from those of open-loop. Special consideration is needed to extract open-loop characteristics from processes that are operating in closed-loop. Such consideration usually leads to the need to apply some persistent and random excitation (maybe even PRBS) to either the set-points or to the MVs of the controller so that the open-loop characteristics are evident in the data.
- *The lack of suitable uncorrelated and rich information within the data.* Adaptive modeling is ineffectual if various cause signals are correlated with each other and/or if certain cause signals are immobile. MPC schemes, by their very nature, frequently lead to MVs being driven to and held constant at constraints and the behavior of FVs is necessarily indeterminate. Lack of excitation in the data can be a serious problem. Long periods of exposure to signals that are essentially just wallowing in noise can ultimately lead to significant deterioration in model quality as the RLS gets distracted by the small correlations that might arise because of common noise between signals.
- *MPC model structures that are large.* Applications in the petrochemical industry can involve many tens of signals. Given the first two factors listed above, it is inconceivable to consider full-blown adaptive MPC operating

continuously against such large processes with all model paths being simultaneously updated as new data arises.
- *The amount of data needed to establish an effective model in real time.* ARX models require far fewer coefficients than FIR models. A control engineer is an impatient person when commissioning new control systems online. Plant managers get irritated if control engineers disrupt operations for too long a period when collecting data for control system identification and design. ARX models can be identified much more quickly than FIR models by the adaptive modeling procedures—as long as the ARX structure is chosen to be appropriate for representation of the process. It is important that the engineer be provided with effective aids to assist in the definition of ARX model structures.

Given the foregoing factors, it is necessary to consider very carefully how adaptive modeling might be effectively exploited to aid the effectiveness of MPC operations. A rule that is strongly recommended is that adaptive modeling as such not be used to continuously update an active controller. Such activity should always be supervised closely by an engineer and should always be short-term. Adaptive modeling should be considered as a tool to enhance the efficiency of MPC modeling projects and as a tool for the maintenance of models that are supporting active controllers.

The following gives example of some of the features have been employed in Connoisseur to allow effective exploitation of adaptive modeling:

- *The definition of segments of model coefficients to be updated.* It is often the case that some aspects of a model are satisfactory and do not need refinement, whereas other portions are empty or do not reflect the process very well. It is of great benefit to be able to mask out the good portions and selectively apply the least-squares analysis so that only the coefficients in the unmasked area are updated. It is then only necessary to have to apply special excitation to the MVs that are associated with the unmasked paths of the model. Further, the masked model can be used to filter out the contributions to the data from MVs and FVs that are masked. Such prewhitening has the advantage of reducing the influence of MV feedback so that the data for the identification of the unmasked parameters has minimized the influence of bias because of closed-loop operation. In the case of an FV signal, this might remove influences that would give rise to biased modeling because of correlation between different FVs. The extent of the effectiveness of the prewhitening filter is of course dependent on the accuracy of the masked model. The use of PRBS excitation might still be required for best effect, if model accuracy is poor.

- *The facility to superimpose PRBS variations on selected MVs of the controller, with a view to generating rich data for the online modeling procedures.* Such disturbance is also useful in injecting independence into the data so that the RLS identifies the open-loop rather than the closed-loop characteristics.
- *The definition of multiple models that are to be referenced by a controller.* The concept of an adaptor model and a tuner model has been introduced in Connoisseur. The adaptor model is the model that is updated by the RLS procedure. The tuner model is the model that is used by the MPC to calculate the control moves. Thus, any initial characterization can be with a model that is updated by the RLS but that is not used by the controller. Only when the indications are that the adaptor model is appropriate would this model be transferred to become the tuner model. Note that the adaptor and tuner models can be the same—implying the continuous updating of model parameters that are being used for control. There may be more than just two models associated with a controller, each for example relating to a different operating region of the process. An infrastructure of linear models might be established to describe a broad range of variation in a nonlinear process, on the basis of piecewise linear approximation. The adaptor provides one means for establishing such a set of models—although the set of models might also be established by analyzing a succession of data gathered in association with the various regions.
- *A set of standard procedures to provide for the automatic management of the online use of models.* An interpretative program language is available that allows the engineer to program the manner in which the multiple models are exploited. For example, it is possible for the prediction error statistics of a complete set of models to be evaluated and for a model to be selected for use with MPC on the basis of best ability to approximate the current operational data. As an alternative, such models can be selected by referring directly to appropriate signals from this data. Another common option is to have a "fail-safe" model that is automatically selected should any problems become apparent with the current tuner model—a safety bolt if mistakes are made.
- *A set of graphical user interfaces to allow the engineer to assess properly the performance of the adaptive modeling and to interact with it.* The engineer must be provided with the appropriate indicators in order to be able to make judgments about model quality and to decide when to initiate and suspend model adaptation. Such indicators are graphical, with trends that play comparisons between model predictions and plant data, and numerical in the form of model error and least-square convergence statistics. The engineer may review the updating model coefficients directly. The engineer may select the adaptor and tuner models and the copy and edit models, and may alter the pace of adaptation by jolting the covariance matrix of the least-squares procedure as described earlier.

OTHER METHODS

Using Radial Basis Functions to Represent Nonlinear Systems

Connoisseur has two primary approaches for dealing with processes that are nonlinear (that is, assuming that the nonlinearity cannot be dealt with via simple transformations of data).

The first is piecewise linearization. The facility is available to store multiple models (known as model sets) that can be established by applying linear identification a number of times across the nonlinear region that is to be modeled.

The second approach is radial basis function (RBF) networks. The main basis for choosing the RBF approach for modeling nonlinear systems is that there is a great deal of compatibility with the approaches described earlier and adopted for linear systems. Thus, the time-series formulation expressed in Equation (2.5) is to some extent retained, the nonlinear formulation taking the more general form

$$\mathbf{y}_{k+1} = \mathbf{f}_N(\mathbf{Y}_k, \mathbf{U}_k, \mathbf{V}_k) \tag{2.9}$$

N is the number of nodes in the RBF description. Row i in Equation (2.9) can be expanded to

$$y(i)_{k+1} = \sum_{j=1}^{N} W_{ij}\Phi(\|z_k - c_j\|) \tag{2.10}$$

with W_{ij} a set of N constant coefficients and c_j a set of N vectors known as centers. In Equation (2.10), the function Φ is commonly chosen to be either Gaussian:

$$\Phi(x) = \exp(-2x^2/N)$$

or logarithmic (thin-plate spline):

$$\Phi(x) = x^2 \log(x)$$

This form of RBF incorporates multiple instances of the vector z_k (N of them), each delivering a radial length to a center c_j within the multidimensional space. Each such center is termed a node of the network. Each radial length is then passed through a nonlinear function and the complete set of outputs is then linearly aggregated in a weighted combination, one weight for each node. The full set of weights is termed the hidden layer.

A major attraction of the RBF approach in the Connoisseur environment is that Equation (2.10) is linear in the coefficients W_{ij}. Thus, if N and $c_j, j = 1 \ldots N$, are known, then the least-squares methods described earlier can be employed to identify the parameters $\mathbf{W}_{ij}$.

All of the Connoisseur tools for the linear case have been upgraded to encompass RBF models, including the capacity to employ adaptive modeling and multiple models.

A most important issue with such networks is the appropriate choice of the centers $\mathbf{c}_j$. The procedure adopted is simply to choose the centers as a set of random numbers. Choice of centers can be more systematic and based on the way that the data clusters in the multidimensional space. There are methods that can analyze such clustering and derive $\mathbf{c}$ and N automatically, for example the "K means" approach. However, these methods impose a substantial burden on any identification procedure, and comparison with the random method has not shown that the additional burden of cluster analysis is justified in terms of improved fidelity in the RBF description. This issue is still being given consideration by the Connoisseur development team.

Given the application of random centers, all that is necessary for choice by the designer in addition to the linear case is the number of nodes N and the type of nonlinear function.

A great deal more care has to be given to the identification of an RBF model than is necessary with a simple linear description. An RBF model derived from a limited set of data and that accurately describes that data can in no sense be assumed to be representative for all conditions. With linear systems, it is reasonable to assume benefits associated with superposition and also that the capability of a model will only degrade gently as the point of operation moves away from the region in which the identification was applied. With RBF models, any such assumptions are invalid. It is essential that models be tested with data that is totally independent of that used for the identification calculations (rather than advisable, as is the case with linear modeling). It is also essential that derived models be used only within the ranges of data variation employed with these calculations. It is also optimistic indeed to expect that a single RBF model might describe all of the nonlinear conditions that might prevail across the full operating range of an industrial process. It makes pragmatic sense to consider that a set of multiple and more regional models might provide a more effective overall representation. The multiple model features of Connoisseur, together with the capacity for adaptive modeling, are therefore of consequence in the nonlinear as well as the linear situation.

Early experiences in the exploitation of RBF models have emphasized the importance of the adaptive utility as a support for the RBF technology. A clear benefit of RBF models is to provide improved prediction capability in the face of process nonlinearity. Such capability can be used for the development of "soft analyzers" that can provide a numerical substitute for expensive instrumentation that is used to monitor process and product parameters. With this aim, the chemical composition of the product of a particular distillation column was very effectively modeled with an RBF model. The model was subsequently installed online to continuously infer the product quality, that is, to provide a soft analyzer in support of the hardware analyzer. After some weeks, the accuracy of the model fell away dramatically. Subsequent diagnosis revealed that a new control system had been installed on the process, without the knowledge of the Connoisseur engineer, and a key parameter, which previously had varied freely, was closely held at a setpoint. This completely altered the process cause-to-effect sensitivities. To recover from this situation, adaptive modeling was initiated and within three to four days the RBF predictions were back on track.

The Partial Least Squares Approach for Model Identification

Partial least squares or projection to latent structures (PLS) is a method of identification that offers certain attractive features, both in providing a more robust identification approach than OLS and in providing a more efficient computation than RLS for use in the offline environment, particularly with large FIR structures. PLS also offers opportunities for exploiting multivariate statistics for condition monitoring in addition to MPC. With reference to Equation (2.6), PLS employs $\mathbf{Z}_k$ and $\mathbf{w}_k$ to fracture matrix α_{ols} into three elements so that

$$\alpha_{pls} = \mathbf{Q}\,\mathbf{B}\,\mathbf{W} \simeq \alpha_{ols} \qquad (2.11)$$

with **B** a diagonal matrix (the inner model).

The diagonal elements of **B** (b_i) and the associated columns of **Q** and rows of **W** are built systematically for $i = 0, 1, \ldots$, etc. At each stage a numerical projection is employed to maximize the explanation of variance in $\mathbf{w}_k$ through the addition of the extra set of parameters.

The procedure stops at some threshold (e.g., when 95% of the variance is explained). Statisticians make much of when to stop, and quite complex cross-validation methods may be employed in an effort to make the choice as to when to stop automatic.

In the extreme (i.e., when i is at maximum)

$$\alpha_{ols} = \alpha_{pls}$$

A major feature of PLS is that it will deal with correlated data because it avoids matrix inversion. Thus PLS will produce answers when OLS cannot. [*Note.* As mentioned earlier, this property is not unique to PLS—for example, recursive least squares (RLS) will also deal with correlated data.]

PLS is also useful in correcting for the situation in which the model structure has been overdefined so that there are too many terms in the model. The clipping of the elements of **B** has the effect of eliminating any redundancy in the model.

Lower energy effects, such as noise, are described by the later additions to **B**, so PLS can be (but is not always) effective in cutting out noise that biases OLS. Thus α_{pls} may be a closer approximation to the true α. Given this fact, there is a case to be made for processing the data twice in order to combine the noise processing benefits of RLS and PLS—that is, if an ARX model is to be identified. RLS will minimize the bias that arises from the presence of noise on the AR terms (PLS will not do this). From the derived ARX model it is possible to replace the actual CV data terms with terms that are predicted from the ARX model and that are, in consequence, filtered. If PLS is then applied, then other more general noise effects can be reduced in the model.

The relationship $\mathbf{t} = \mathbf{W}\,\mathbf{z}_k$ provides a transformation from the cause variables to inner variables (scores or latent variables, LVs). The dimension of **t** may be much smaller than that of $\mathbf{z}_k$.

It is this property of PLS that gives opportunity for process condition monitoring. The vector **z** may involve many variables from a process, and many of these may be correlated with each other. It is often the case that the required number of LVs is quite small. Thus rather than having to consider the "condition" of the process by having to interpret the behavior of all of the signals in **z**, the problem may be considerably simplified by interpreting the behavior of the LVs **t**.

SIMULATED CASE STUDY ON A FLUID CATALYTIC CRACKING UNIT

Description of the FCCU

A fluid catalytic cracking unit or FCCU is an important economic unit in refining operations. It typically receives several different heavy feedstocks from other refinery units and cracks these streams to produce lighter, more valuable components that are eventually blended into gasoline and other products. The particular Model IV unit described by McFarlane et al. (1993) is illustrated in Figure 2.1. The principal feed to the unit is gas oil, but heavier diesel and wash oil streams also contribute to the total feed stream. Fresh feed is preheated in a heat exchanger and furnace and then passed to the riser, where it is mixed with hot, regenerated catalyst from the regenerator. Slurry from the main fractionator bottoms is also recycled to the riser. The hot catalyst provides the heat necessary for the endothermic cracking reactions. The gaseous cracked products are passed to the main fractionator for separation. Wet gas off the top of the main fractionator is elevated to the pressure of the lights end plant by the wet gas compressor. Further separation of light components occurs in this light ends separation section.

The Exploitation of Adaptive Modeling

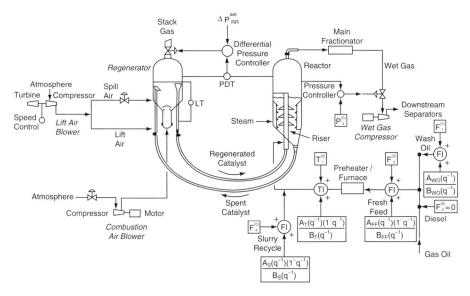

Figure 2.1 Schematic diagram of the fluid catalytic cracking unit.

As a result of the cracking process, a carbonaceous material, coke, is deposited on the surface of the catalyst, which depletes its catalytic property. For this reason, spent catalyst is recycled to the regenerator where it is mixed with air in a fluidized bed for regeneration of its catalytic properties. Oxygen reacts with the deposited coke to produce carbon monoxide and carbon dioxide. Air is pumped to the regenerator with a high-capacity combustion air blower and a smaller lift air blower. As well as contributing to the combustion process, air from the lift air blower assists with catalyst circulation. Complete details of the mechanistic simulation model for this particular model IV FCCU can be found in McFarlane et al. (1993).

The model is run at about 60 times faster than real time in order to maximize the efficiency of the studies that are described below. Thus the 2–second update interval referenced for the models that are derived corresponds in actuality to about 2 minutes on a real FCCU. Further, and to simulate realistic disturbance conditions, various artificial noise signals [e.g., autoregressive moving average (ARMA) signals] are superimposed onto some of variables.

The model predictive controller that is to be considered has 4 CVs and 3 MVs. The CVs are:

- Riser temperature (M308): to be controlled to a setpoint
- Regenerator bed temperature (M314): to be maintained within the range 1250 to 1272°F
- Stack gas oxygen concentration (M317): to be controlled to a setpoint

- Wet Gas Suction Valve position (M334): to be maintained within the range 0.2% to 0.95% open

The MVs are:

- Setpoint to the wash oil flow controller (A100): to be maintained within the range 0 to 17 lb/s
- Setpoint to the fresh feed flow controller (A102): to be maintained within the range 0 to 144lb/s
- Set point to the lift air flow controller (A108): to be maintained within the range 0 to 200 lb/s

The associated codes for each signal (e.g., M308, A100) are the Connoisseur identities that will be evident on the various figures presented later.

Three exercises are described:

- *The application of step tests to generate data to identify an initial model from which an initial working controller may be developed.* This exercise serves to illustrate the character of the process and provides the basis for selection of model structure and initial characterization of model coefficients.
- *The use of adaptive modeling to determine coefficients for model paths not characterized by the initial identification.* This is done with MPC in action, but without using MV A108 (the lift air flow controller setpoint). It is presumed that no initial model coefficients are available for all of the paths from A108 into the CVs. The purpose of the exercise is to establish these coefficients while the remaining two MVs are in active use with the controller. In this example an FIR model is considered.
- *The use of adaptive modeling to identify a complete new model.* This is also done with MPC in action, but this time with all MVs in active use. The purpose is to establish and employ a new model with the controller, but without having to disable active control at any stage. In this example an ARX model is considered.

Application of Step Tests

A series of steps are applied to the MVs as illustrated in Figure 2.2. This figure shows a set of 8 trends in total, these being allocated to horizontal windows of equal dimension. Each trend is scaled to fit its window. To the left of each window are the details for the associated signal, including name and units text and the minimum and maximum values of the trend that is on display. The values in white are the signal values that correspond to the intersections with the vertical axis of the white crossed lines (a cursor). Beneath the x-axis is information that gives the times corresponding to the leftmost and rightmost extremes, the time at the cursor and the time span.

The Exploitation of Adaptive Modeling 79

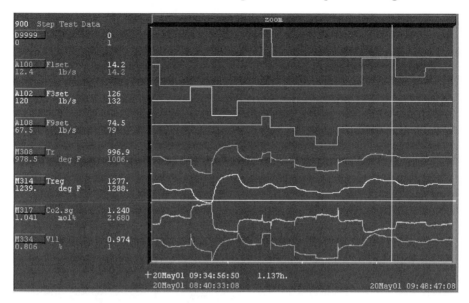

Figure 2.2 Step test data for initial identification.

Figure 2.2 illustrates values for the CVs and MVs reviewed above, together with the signal D9999, which is used as a mask for data that is not to be included for identification. This mask is set for a short period within the second quartile of the data. Inspection of the signal M334, the valve position, shows it to be saturated at 100% value across this period. The mask prevents data associated with this nonlinear effect from being employed in the identification. The MVs are stepped in turn, each being altered 3 or 4 times within 5 to 10% of full range. The corresponding responses of the CVs are clear and the intervals between change in the MVs are sufficient for the CVs to settle.

Figure 2.3 illustrates the structure settled on for the initial identification. The Prediction Interval of the model is set at 2 sec. The model is chosen with 0 as the order of dynamics, which implies either an FIR or a steady-state form for the model. In such a case, the "delay spread" shown in Figure 2.3, determines the number of coefficients for each MV (0 implies steady-state). It can be seen that A100 is allocated a delay-spread of 100 units (each unit being 2 sec), A102 a delay-spread of 150, and A108 a delay-spread of 100. These choices are made by reference to Figure 2.2 and are the observed (very) approximate maximum times for each parameter to reach steady state following the changes in the MVs.

Figure 2.4 presents the results of the identification. The smooth lines are the model predictions and they are superimposed on the associated CVs. The model is initialized with the data at the far left-hand side and thereafter only makes reference to the MVs in order to progress its predictions (except at the point at which the

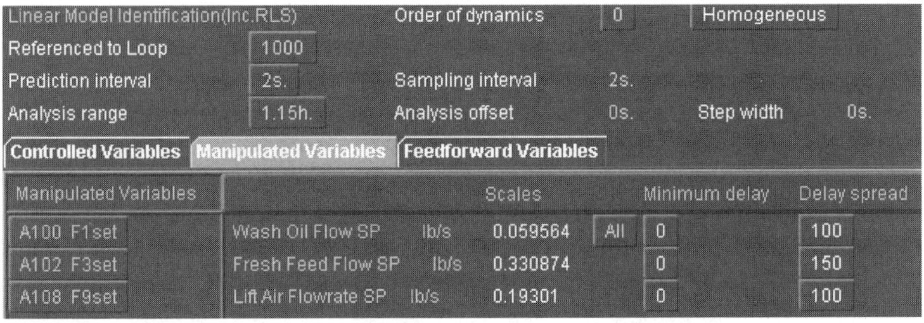

Techniques for Adaptive Control

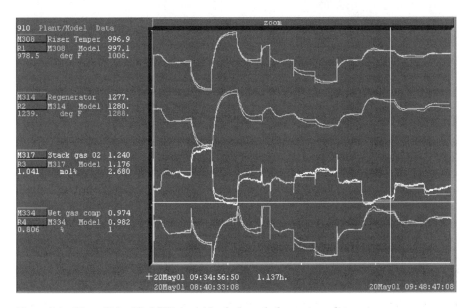

Figure 2.3   Structure of FIR model.

Figure 2.4   Play of identified FIR model back through the step test data.

masked data is encountered, at which stage the model is once more initialized with CV data). To be strictly thorough, the model should be assessed against data that has not been used in the identification calculation—however, for instances where the data is richly excited and where the number of data points is greatly in excess of the number of coefficients in the model, as is the case here, most often the judgment may be made that the cross-validation exercise is not necessary. Figure 2.4 shows that the variations in the data are largely well defined by the model. Discrepancies are present because the process is not linear and because of the superimposed noise, some of which is of nonstationary character. It is also clear that the process is characterized by a mix of very fast and very slow effects. Some of the CVs exhibit sharp initial changes of spike character, particularly following a change in the MV A108, to be followed by slow transients that take as long as 5 minutes to settle. It is necessary that

The Exploitation of Adaptive Modeling 81

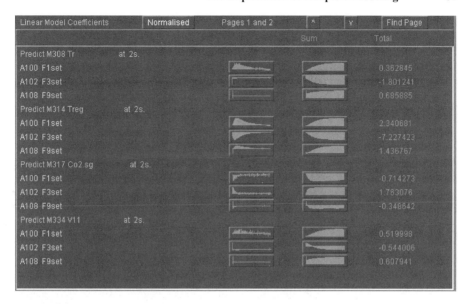

Figure 2.5 FIR model coefficients.

the sampling and prediction associated with the model be at 2 sec in order to pick up the fast transients. The number of FIR terms needs to be ≥ 100 in order to pick up the slow transients. However, it is clear that the derived model provides a largely accurate description of the majority of the variation in the data.

Connoisseur supports a wealth of statistical detail and residual graphic displays to provide the information for the design engineer to effectively evaluate and refine the identified model. Discussion of these aspects is beyond the scope of the chapter.

Figure 2.5 presents the coefficients that are associated with the various paths of the model. Each cause-to-effect path has three associated pieces of information:

- A histogram of the actual FIR coefficients
- A histogram of the cumulative sum of these coefficients (hence forming the finite-step response or FSR)
- A sum total of the FIR coefficients—which is the steady-state gain for the path

The FSR characteristics in particular emphasize the complex character of the process under study. Some of the paths (e.g., A108 to M308 and M317) are dominated by the short term, some by the long term (e.g., all paths from A100), and some exhibit non-minimum-phase behavior, particularly the path A102 to M334. This mix of issues is typical of situations encountered in the process industry.

Figure 2.6 presents the coefficients of the associated **G**(*s*) transfer function. Each block presents, for all of the model paths, one of the transfer function parameters. The five blocks present the gain, the two denominator time constants, the numerator time constant and the time-delay, respectively, for each of the model paths (see Equation 2.2). If the second time constant is zero, this implies first-order rather than second-order dynamics. If both the first and the second denominator time

K Gain Mx.	A100	A102	A108
M308	2.579362	-2.305067	1.504684
M314	7.072654	-3.931378	1.339769
M317	-0.285234	0.126744	-0.042965
M334	0.034566	-0.006509	0.012471

T1 Mx.(s.)	A100	A102	A108
M308	-0.770356	0.887712	11.81406
M314	-0.815227	7.586789	9.590652
M317	17.57907	10.39534	0.991811
M334	-0.807711	7.246327	7.054969

T2 Mx.(s. or rad/s.)	A100	A102	A108
M308	-0.025858	55.31296	0
M314	-0.026949	58.67138	43.15207
M317	65.89915	79.8167	6.902059
M334	-0.029639	56.722	0

To Numerator Time Constant Mx.(s.)	A100	A102	A108
M308	19.39112	18.65293	0
M314	21.5234	1.04171	-9.923756
M317	99.11116	68.57043	-15.31535
M334	8.220025	-72.2298	0

Td Delay Mx.(s., d=Td/T)	A100	A102	A108
M308	7.095631	0.623617	2
M314	9.188572	2	0.792154
M317	2.373477	2.679097	2
M334	5.623718	0	2

Figure 2.6 Transfer function parameters.

constants are negative, this implies an underdamped response—in which case the first constant is the damping ratio and the second constant is the natural frequency. The MV A100 in particular exhibits underdamped characteristics. Notice the negative numerator time constant that corresponds to the paths that are nonminimum phase. The information within Figure 2.6 is very useful in helping an engineer to assess the capability of a model in terms of characteristics that relate directly to the way the process is understood to respond. It is also of great value in helping to assess the character/structure of an ARX model should one be desired—as is discussed in more detail later.

Adaptive Modeling of the Coefficients of a Single Model Path

The model just described has been used to establish a model predictive controller, with the outline performance specification reviewed earlier in the description of the FCCU. The details of this controller are not within the scope of this chapter, and information about the controller is presented here only insofar as is necessary to explain behavior that is relevant to the modeling exercises. The control system exploits quadratic programming (QP) in order to manage constraints, setpoints, etc. It updates every 2 sec on the basis of a prediction horizon of 100 sec. For the following considerations, this controller may be assumed to be active and online throughout, managing the process continuously while the adaptive modeling procedures are in progress.

The derived FIR model gives rise to a control system that has very satisfactory behavior in all respects. There is therefore no need to exploit adaptive modeling to update model coefficients. To implant such a need, the coefficients associated with all paths from the MV A108 are set to zero and the control system is restricted to operate with only the two MVs A101 and A102. In this example, the adaptive modeling exercise is therefore to characterize the paths from A108 while active control is in force to manage the process using A101 and A102. Thus in actuality, in this example, the signal A108 is being treated as a feedforward variable.

Figure 2.7 illustrates 15 minutes of process operation. The CVs M308 and M317 are being held to setpoint and the CVs M314 and M334 are being held within a range. Masks are defined so that only the coefficients in the paths from A308 may update via the adaptive modeling. The adaptor is initialized in the early phase of the illustrated operation. This amounts to setting a high value in the covariance matrix of the least squares procedure that operates online (essentially **P** in Equation 2.8). The MV A308 is subsequently moved manually on four occasions up and down between 65 and 75 lb/s and Figure 2.7 illustrates the control system responding to maintain its objectives in the face of the disturbance impact from A308. The RLS procedure (see Equation 2.8) is updated at each model sampling instant and, eventually, once enough data has been processed, the covariance matrix converges to very small values.

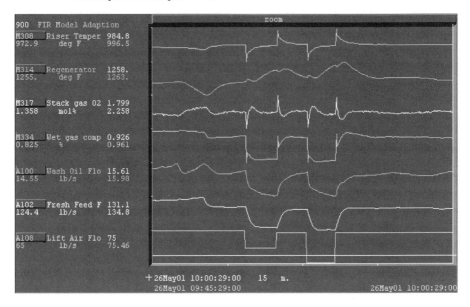

Figure 2.7   Adaptive modeling of a single FIR model path: test sequence.

Figure 2.8 presents the predictions that arise from the adapting model for two of the model CVs—M308 and M314. The bright traces represent the predictions superimposed on the CVs. These predictions are calculated and displayed in real time as the adaptive modeling progresses. The calculation is carried out with reference to the CVs at and before an instant in time prior to the current instant. The model is then iterated for the number of prediction steps necessary to bring the predicted value from the earlier instant up to current time. Only MVs are referenced during the intermediate steps. It is absolutely critical that sufficient prediction steps be employed that the engineer is able to properly assess the prediction capability of the model. In this example, 20 prediction stages are used, amounting to 40 sec in total. Too few prediction steps always leads to false confidence in the model with bad results in consequence once the model is employed for control. With reference to Figure 2.8, the character of behavior of the predictions is very different for M308 and M314. Referring back to Figure 2.2, it can be seen that M314 is very smooth in its response to all MVs, with little evidence of the fast transients that disrupt the other CVs, whereas M308 is subject to spike transients. These spikes reflect in the coefficients of the adapting model as these coefficients are updating. Following the four step changes in A108 and the subsequent period of settling, confidence indicators such as the convergence of the covariance matrix and the stability and accuracy of prediction suggest that the newly established coefficients provide realistic representation of the process.

Figure 2.9 illustrates the performance of the derived model, with the adaptive modeling disabled following the identification exercise described above. A108 is

The Exploitation of Adaptive Modeling 85

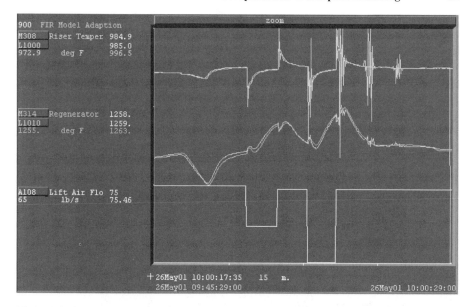

Figure 2.8   Adaptive modeling of a single FIR model path: evolving real-time predictions.

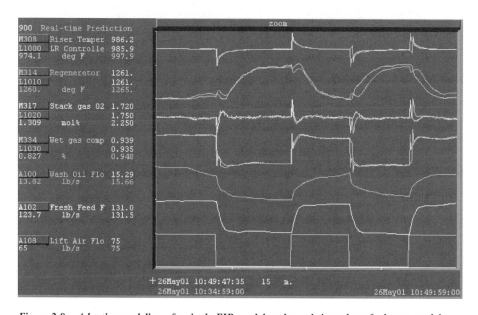

Figure 2.9   Adaptive modeling of a single FIR model path: real-time play of adaptor model.

seen to be moved up and then down on two occasions. For the first time, the predicted values are generated from the newly derived model as before and are seen to track with reasonable accuracy. For the second time, the predicted values are generated from the model that has the coefficients associated with A108 set to zero.

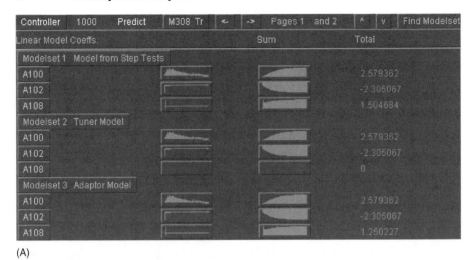

Figure 2.10   (A–D) Adaptive modeling of a single FIR model path: derived coefficients.

The relative deficiency of the latter is quite apparent, particularly in M314 for the long transients and in the other CVs for the short transients.

Finally, Figure 2.10 presents the derived model coefficients. There are four sections to the figure, with each relating to a particular CV. In each section, the model coefficients are presented for three model sets. The first model set is that established from the initial step tests. It is not used in the real-time exercise described above but is present for reference. The second model set is that that is used by the controller (i.e., the tuner model set). Note the zero coefficients associated with MV A108. The third model set is the one that has been updated online—although only the paths associated with

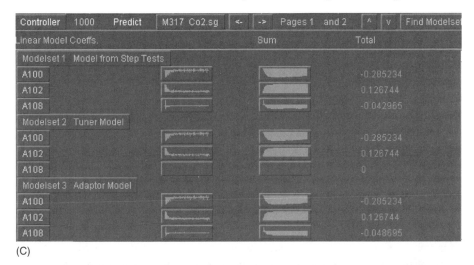

(C)

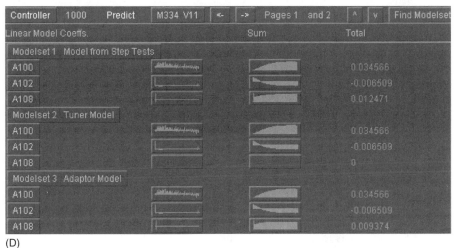

(D)

Figure 2.10 (continued)

A108. The other paths are preset to be the same as those of model set 1, prior to the commencement of adaptive modeling. Comparing the paths for A108 between model sets 1 and 3 shows that the coefficients derived online are very similar to those derived from the original step test data. The patterns of numbers within the histograms are almost identical and the sum totals of these numbers are very close. Note that this accuracy of identification is achieved despite the closed-loop control operations that prevailed throughout. This is possible because of the availability of the existing model for the paths from A101 and A102. This known portion of the model is used as a filter, each time a new model calculation is made, so that the effects of A101 and A102 on the CVs are removed from the data before any calculation associated with

88   Techniques for Adaptive Control

A108 is carried out. This aids to reduce any bias that might arise because of the closed-loop operations.

Adaptive Modeling of a Complete Model Structure

An example is now given that uses an ARX formulation rather than FIR. First, an ARX model is established from the initial step data of Figure 2.2. This ARX model is then used as the basis for a model predictive controller. The exercise is then to establish a new model with coefficients that describe all paths, to do this during the course of controller operation, and to subsequently employ the newly identified model for control. This is therefore an example of real-time adaptive control in a multivariable domain.

Inspection of the transfer function parameters of Figure 2.6 indicates that the minimum time delay present is 0 and the maximum is about 10 sec or five prediction steps of the FIR model. The maximum time constant is between 1 and 1.5 minutes, and there is a complex mix of first- and second-order transients that are variously damped, underdamped, and nonminimum phase. In consequence of all of this, an ARX structure is selected to have second-order dynamics overall, with a heterogeneous format (see Equation 2.5). The "delay spread" or maximum time delay for each MV is set to 10, in order to err on the safe side. These selections are indicated in Figure 2.11. The final selection arrived at is a mix of trial-and-error and judgment based on the foregoing information.

Figure 2.12 indicates some of the identified coefficients of the derived ARX model (for the prediction of M308 and M314). Note the terms associated with the CVs, which are of course absent in the FIR case. Also note the greatly reduced number of terms associated with each MV (i.e., for 100 or more down to 10). These results rely heavily on the filtering mechanism of the RLS procedure that is described above in association with Equations (2.8). In fact, to achieve convergence to the degree shown, the RLS analysis has to be recycled through the data some 20 times.

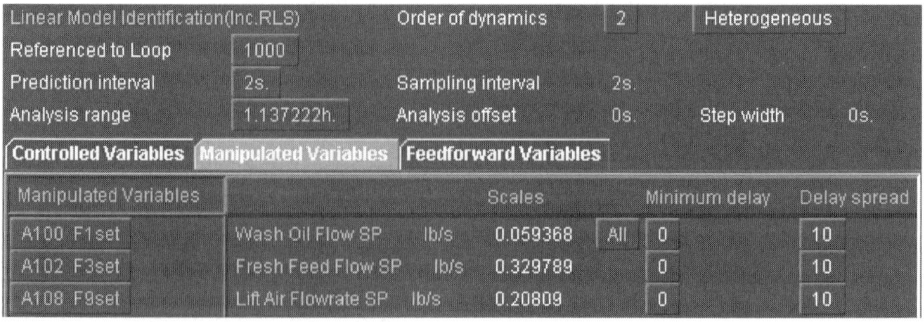

Linear Model Identification(Inc.RLS)		Order of dynamics	2	Heterogeneous	
Referenced to Loop	1000				
Prediction interval	2s.	Sampling interval	2s.		
Analysis range	1.137222h.	Analysis offset	0s.	Step width	0s.
Controlled Variables	**Manipulated Variables**	**Feedforward Variables**			
Manipulated Variables		Scales		Minimum delay	Delay spread
A100 F1set	Wash Oil Flow SP	lb/s	0.059368 All	0	10
A102 F3set	Fresh Feed Flow SP	lb/s	0.329789	0	10
A108 F9set	Lift Air Flowrate SP	lb/s	0.20809	0	10

Figure 2.11   Structure for ARX model.

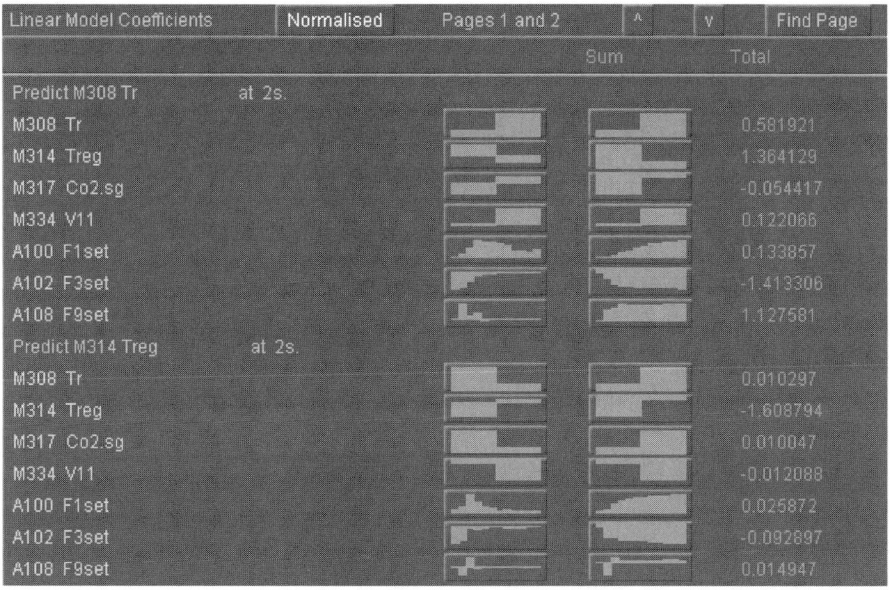

The Exploitation of Adaptive Modeling

Linear Model Coefficients	Normalised	Pages 1 and 2	^	v	Find Page
			Sum	Total	
Predict M308 Tr	at 2s.				
M308 Tr					0.581921
M314 Treg					1.364129
M317 Co2.sg					-0.054417
M334 V11					0.122066
A100 F1set					0.133857
A102 F3set					-1.413306
A108 F9set					1.127581
Predict M314 Treg	at 2s.				
M308 Tr					0.010297
M314 Treg					-1.608794
M317 Co2.sg					0.010047
M334 V11					-0.012088
A100 F1set					0.025872
A102 F3set					-0.092697
A108 F9set					0.014947

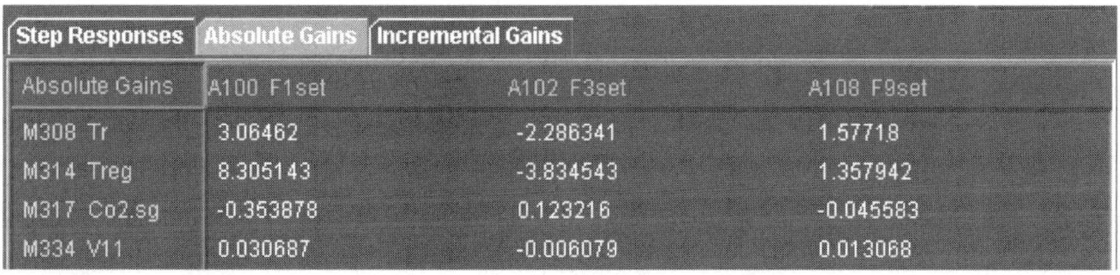

Figure 2.12   Coefficients of identified ARX model.

Step Responses	Absolute Gains	Incremental Gains	
Absolute Gains	A100 F1set	A102 F3set	A108 F9set
M308 Tr	3.06462	-2.286341	1.57718
M314 Treg	8.305143	-3.834543	1.357942
M317 Co2.sg	-0.353878	0.123216	-0.045583
M334 V11	0.030687	-0.006079	0.013068

Figure 2.13   Steady-state gains of identified ARX model.

Step Responses	Absolute Gains	Incremental Gains	
Step Responses	A100 F1set	A102 F3set	A108 F9set
M308 Tr			
M314 Treg			
M317 Co2.sg			
M334 V11			

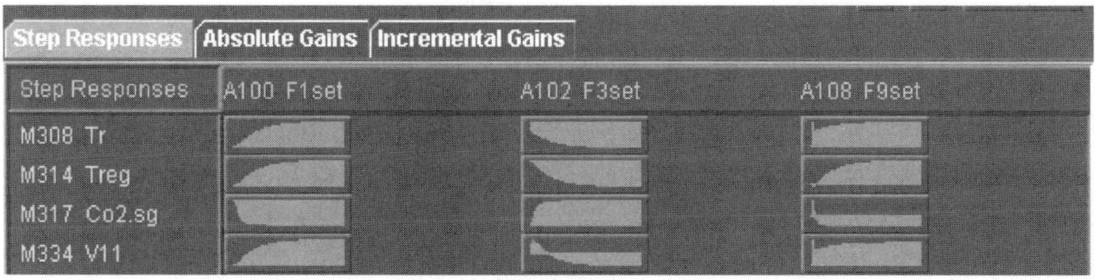

Figure 2.14   Unit step responses of identified ARX model.

Figures 2.13 and 2.14 illustrate the steady-state gains and the unit step-responses for each of the cause-to-effect paths of the derived model. Comparison of the gains with those of Figure 2.6 shows good correspondence with the FIR case, although some of the gains associated with the MV A100 are up to 20% larger in the ARX model. The step responses are closely similar although derived by very different means—that is, in the FIR case by summing the coefficients and in the ARX case by actually applying a unit step and subsequently iterating the model to a steady state.

Figure 2.15 presents the play of the ARX model against the plant data. These results can be seen to be very similar to that of Figure 2.4 for the FIR case. There is therefore strong confidence that a control system designed from the ARX model will perform equally as well as that using the equivalent FIR model—which proves to be the case in practice. Note that a controller based on an ARX model has operational advantages—particularly in dealing with the elimination of impact from unmeasured disturbances.

Figure 2.16 illustrates 28 minutes of the operation of such an ARX-based model predictive controller. In this period, a number of setpoint changes are made in relation to both M308 and M317. These CVs respond appropriately to these demands while all other variables are managed properly within constraints—that is, the control system behaves according to requirements. Further, adaptive modeling is in progress for most of the time that is displayed.

Figure 2.16 shows a display that is of value in overseeing the operation of adaptive modeling. This is a very busy display—necessarily so because of the number and variety of parameters that need to be monitored by the engineer in order to make judgments as to the quality of the progressing model. The display has three trend

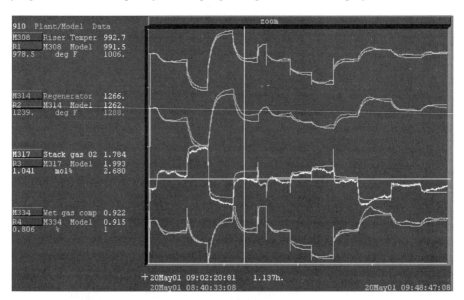

Figure 2.15 Play of identified ARX model against plant data.

The Exploitation of Adaptive Modeling 91

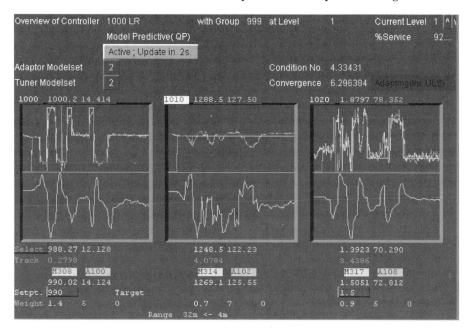

Figure 2.16 Adaptive modeling of a complete ARX model: real-time overview of adaptor operation.

windows, each divided horizontally into two segments. The three top segments show the trends for the CVs M308, M314, and M317, respectively, together with associated setpoint values and associated predictions. The bottom three segments show the trends for the MVs A100, A101, and A108, respectively. The display pans left and right to present further windows relating to other CVs and MVs of the model. Many numbers are presented, indicating the current values of the signals and setpoints, the minimum and maximum values of the trends, and the time range of the trends. The following information is of value in support of adaptive modeling:

- *Clear indication of the general excitation of plant signals.* Parameters that are in constraint will be poorly modeled because of insufficient variability.
- *The degree to which the RLS procedure has converged.* The trace of the covariance matrix **P** is presented against the text "Convergence." The smaller this value, the greater is the confidence that the established model might be appropriate. *Note*: No model should be considered of value if the "Convergence" is large (i.e., <10). A small value of "Convergence" means that the identification has progressed to a stage at which it is appropriate to assess the model using other indicators.
- *The accuracy with which the prediction trends track the associated CVs.* This can be judged by eye from the trends. In addition, each trend window has

an associated "track" value, which is presented on the second line beneath the window. Smaller values indicate better prediction tracking.
- *Indication of the model sets that are in use (tuner for the controller, adaptor for the adaptive modeling).* If the two are the same, the implication is that adaptive control is in operation in the sense that the model being used for control is continuously updating on the basis of new information received.
- *Indication of the condition number of the model that is in use for control.* A large condition number suggests that the controller does not have the degrees of freedom to manage the issues with which it is faced. This may be an interpretation made of a model that is accurate but it is also very commonly an indication that a model is defective.
- *Visibility of the identified model coefficients in order to assess the stability of the updating values and the overall integrity of the model.* Figure 2.17 shows a companion display to that of Figure 2.16 that presents the latest coefficients, updating in real time, for all of the unmasked paths of the identified model.

The preceding factors emphasize that the successful exploitation of adaptive modeling involves significant judgment and skill and interaction from the engineer.

Figure 2.18 presents all of the trend information shown in Figure 2.16 within a single trend display, together with information concerning M334, the fourth CV. In this case the setpoints appear as a series of steps.

The CV M308 is shown in the top segment of the trend. The setpoint for M308 (S1000) is varied between 990 and 1000°F, with eight changes in all across the span.Similarly, the CV M317 is shown in the third segment of the trend (from

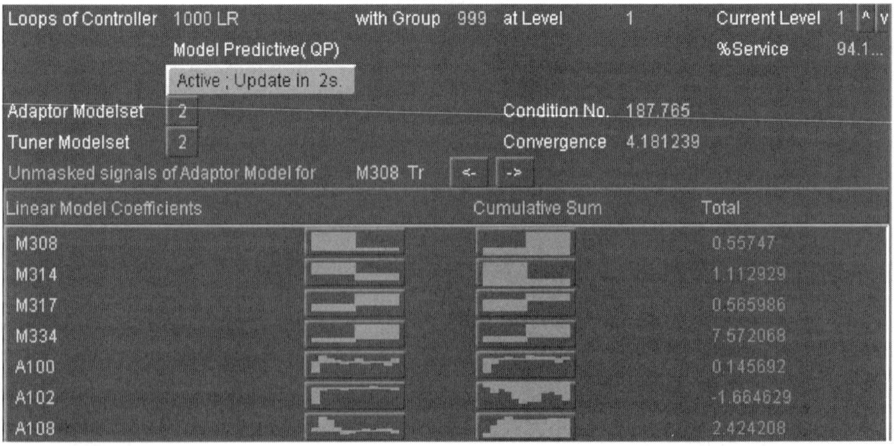

Figure 2.17 Adaptive modeling of a complete ARX model: real-time display of updating coefficients.

The Exploitation of Adaptive Modeling 93

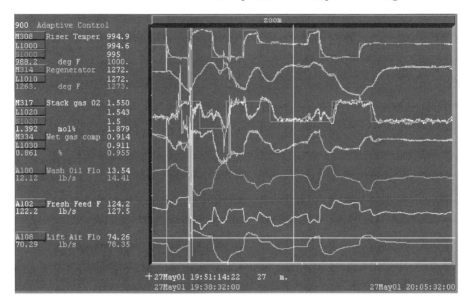

Figure 2.18 Adaptive modeling of a complete ARX model: updating predictions from adaptor model.

the top). The setpoint for M317 (S1020) is varied between 1.4 and 1.8 mol%, with 6 changes in all across the span. These changes are made to excite the process sufficiently so that the adaptive modeling is presented with data that are acceptable for identification.

The adaptive modeling is switched on just prior to the first change in S1000 (at the left of the display) from 995 to 990°F. The sharp spike in the prediction trends is an indication of this event. Thereafter setpoint changes up to the vertical axis of the white cursor arise in sequence as follows:

- (a) S1000; 995 to 990: After this change the model predictions appear to immediately track well but the convergence value is very high.
- (b) S1020 1.5 to 1.8: Predictions for M317 and M334 go wild, with large offset.
- (c) S1000 990 to 1000: Predictions go very wild on all variables for a short period and thereafter return to stability and reasonable tracking.
- (d) S1020 1.8 to 1.4: Setpoint change is only partially followed because both M314 and M334 ride into upper constraints.
- (e) S1000 1000 to 990: Constraints free up. M317 recovers to the setpoint. There is some volatility in the predictions, particularly for M314 and M317.
- (f) S1020 1.4 to 1.8: M317 responds smartly this time. There are large spikes to be seen in the predictions for M308 and M334.
- (g) S1000 990 to 1000: Prediction tracking is consistent, with no further large deviations of consequence. This is also the case for all subsequent setpoint changes.

94 Techniques for Adaptive Control

(h) S1000 1000 to 995: Excursion of M317 from setpoint arises because of M334 being managed at high constraint.
(i) S1020 1.8 to 1.5: Normal response.
(j) Cursor: At the instant corresponding to the cursor, the controller is switched to employ the new model, although the adaptive modeling continues to function. This choice is made on the basis of the observed acceptable tracking, a low convergence figure, and reasonable stability and sense in the model coefficients. The various subsequent setpoint changes indicate a character of responses that is indistinguishable by eye from that arising in the earlier phase, with the original model being employed.

Figure 2.19 shows a comparison of the coefficients associated with the prediction of M308, between those of the initial model (model set 1) and of the final model (model set 2). There is a general approximate correspondence, apart from the coefficients for M317 and A100, which appear very different. Such discrepancy arises with respect to all of the predicted CVs (not just M308) and suggests that aspects of the new model should be investigated in more depth in order to retain confidence. Figure 2.20 shows the final model being played against the data once again, but this time in the offline manner, with the model being initialized once only with data to the left (as in Figure 2.4). This is a much more testing way of assessing the predictive capability of the model than shown in Figure 2.17 (in which reference to the CVs is constantly reset prior to predicting forward by 20 steps only). In Figure 2.20, the model tracks all CVs

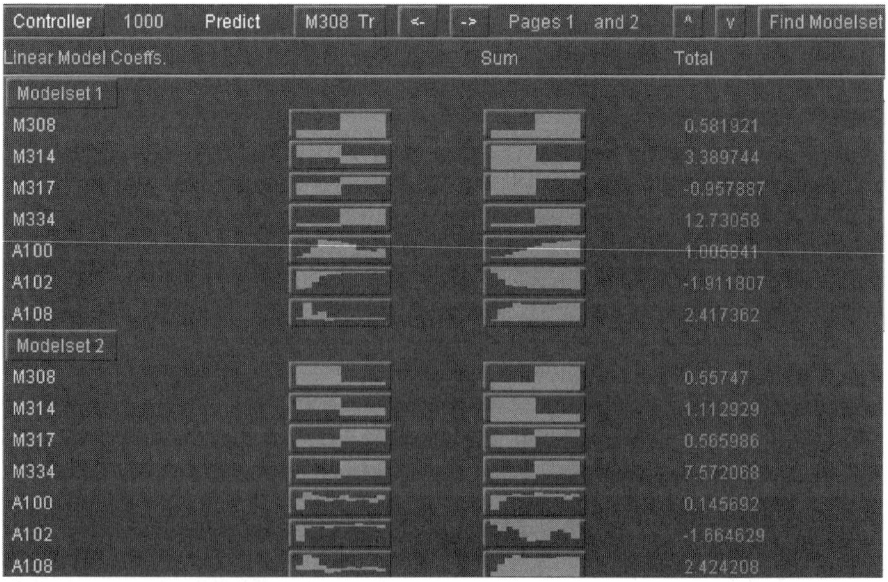

Figure 2.19 Adaptive modeling of a complete ARX model: Comparison between original and updating coefficients.

The Exploitation of Adaptive Modeling 95

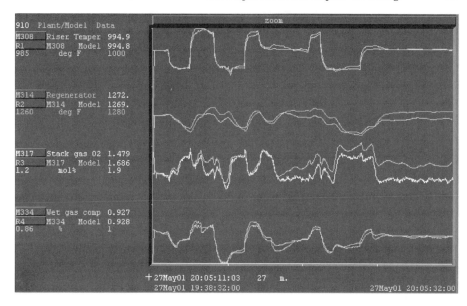

Figure 2.20 Adaptive modeling of a complete ARX model: Play of adaptor ARX model through plant data.

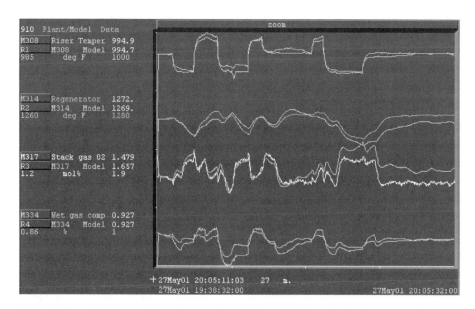

Figure 2.21 Adaptive modeling of a complete ARX model: Play of initial ARX model through plant data.

with reasonable accuracy (given the random and nonstationary disturbances that are being applied). It would appear that the relationship for M314 is the least accurate, although the general pattern of variation is still quite closely maintained. Figure 2.21 shows the initial model being played against this data in the same way. It is apparent that the final model represents this data more accurately in certain respects, particularly in describing the variation of M334. The conclusion is therefore that the final model is a very satisfactory representation, which has improved on the original and that the discrepancy between the coefficients is to advantage.

CONCLUSION

The case study described in this article illustrates the capability of the adaptive modeling procedures that have been implemented within the MPC product Connoisseur. This capability is an important offering for the ongoing maintenance and development of model predictive controllers that are installed on industrial processes. The character of the results described here has been reflected in many of the applications in which Connoisseur-based control systems have been applied in industry.

It is appropriate to express a level of caution in these conclusions. The proper use of adaptive modeling must involve the skill and judgment of a professional engineer. The real-time modeling procedures can present false if answers if not used properly. Data must be rich with variation and the control system must be nonlinear in its mode of working when such data are collected if full adaptive control is to be exploited. For a model predictive controller the latter mode is actually quite normal, because such control is required to drive the process to constraint boundaries. In the exercise described in the section entitled "Adaptive Modeling of a Complete Model Structure," some of the setpoint changes made were deliberately chosen so that the constraint management properties of the controller would be invoked, thereby ensuring that the relationships between the CVs and MVs in the feedback path are not consistent. In such a circumstance, the only consistent relationships between the CVs and MVs are the forward (i.e., open-loop) paths of the process. An alternative approach to dealing with the issues of closed-loop identification is to employ PRBS disturbances to "ripple" the CVs and MVs of the controller in random fashion.

Note that this latter requirement does not pertain for the single-path modeling exercise described in the section entitled "Adaptive Modeling of the Coefficients of a Single Model Path." In this case the whole exercise is more straightforward and less subject to any complexities that might arise because of the scale of the control management problem. The identification issues may be considered on a path-by-path basis. MVs can be taken in turn, out of the working multivariable controller, and be treated temporarily as FVs for the purpose of remodeling. It is not necessary to switch off a multivariable controller because one of its MVs has been taken out of play. Once an effective description for the particular path has been established, the MV can be reinstalled with the working controller.

In conclusion, therefore, adaptive modeling is an effective tool for the maintenance and refinement of model predictive controllers. It is a facility that, when used properly by the engineer, will enable the continuing performance of control systems to be maintained at peak, to the quality of the original installation. A serious weakness of MPC schemes has been the deterioration in quality that comes about slowly, perhaps over many months, because the characteristics of processes drift and models become inaccurate. Adaptive modeling is an effective address to overcome such deterioration in an efficient manner.

REFERENCES

Åström, K. J., and B. Wittenmark (1989). *Adaptive Control*. Addison Wesley, Reading, MA.

Bierman, E. J. (1976). "Measurement Updating Using the U-D Factorization," *Automatica*, 12.

Clarke, D. W., and P. Gawthrop (1975). "Self Tuning Controllers," *Proc. IEE*, **122**(9), 929–934.

Geladi, P. (1988). "Notes on the History and Nature of Partial Least Squares (PLS) Modeling," *J. Chemometrics* **2**, 231.

Hastings, J. R., and M. W. Sage (1969). "Recursive Generalized-Least-Squares Procedure for Online Identification of Process Parameters," *Proc. IEE* **116**(12), 2057.

Haykin, S. (1994). *Neural Networks*. Macmillan, New York.

Hesketh, T., and D. J. Sandoz (1987). "Application of a Multivariable Adaptive Controller," *Proc. ACC*.

Kuo, B. (1982). *Automatic Control Systems*. Prentice-Hall, Englewood Cliffs, NJ.

Ljung, L. (1987). *System Identification, Theory for the User*. Prentice-Hall, Englewood Cliffs, NJ.

McFarlane R. C., R. C. Reineman, J. F. Bartee, and C. Georgakis (1993). "Dynamic Simulator for a Model IV Fluid Catalytic Cracking Unit," *Comp. Chem. Eng.* **17**, 275.

Qin, S. J., and T. A. Badgewell (1997). "An Overview of Industrial Model Predictive Control Technology" Internet: www.che.utexas.edu/~qin/cpcv/cpcv14.html.

Sandoz, D. J. (1984). "CAD for the Design and Evaluation of Industrial Control Systems," *Proc. IEE* **131**(4).

Sandoz, D. J, B. Lennox, P. R. Goulding, T. Kurth, M. J. Desforges, I. S. Woolley (1999). "Innovation in Industrial Model Predictive Control," *IEE Computing Control J.* **10**(5).

Sandoz, D. J., and B. H. Swanick (1972). "A Recursive Least Squares Approach to the Online Adaptive Control Problem," *Int. J. Control*, 16.

Sandoz, D. J., and O. Wong (1979). "Design of Hierarchical Computer Control Systems for Industrial Plant," *Proc. IEE* **125**(11).

3

ADAPTIVE PREDICTIVE REGULATORY CONTROL WITH BRAINWAVE

Mihai Huzmezan, William A. Gough, and Guy A. Dumont

Industry has primarily used the proportional–integral–derivative (PID) controller to perform regulatory control of processes. In practice such controllers are often poorly tuned, resulting in unsatisfactory control. At the same time the PID controller displays poor robustness when dealing with processes exhibiting dead time, even when augmented with a Smith predictor.

Model-based regulators have been developed as alternatives to handle difficult process control problems. Typically such methods are not easy to use and require expertise to apply. To overcome such issues a new approach has been developed based on modeling the process response using a Laguerre function series approximation method. The Laguerre approach enables the controller to automatically model the process response, reducing significantly the expertise required by the user. This approach is further connected with an easy-to-use model-based predictive controller providing control solutions for a number of applications.

This chapter provides the reader with an overview on indirect adaptive predictive control embedded in the commercial adaptive controller BrainWave produced by Universal Dynamics Technologies Inc. This technology is recommended for the transition from classic to advanced control when dealing with process challenges such as deadtime, long time constants, integrating characteristics, or significant nonlinearities.

Advanced control of a complex industrial system involves three major steps: model selection, parameter estimation, and controller design. Quite often a model based on first principles is difficult to obtain. Therefore, in practice, an approach based on an input/output model is more appealing. Although linear models are only valid over a reduced range, they are still commonly selected for control. For a number of processes the nonlinearities are such that a control based on a fixed linear approximation cannot be satisfactory, hence the motivation for an adaptive scheme.

In this chapter we discuss a novel approach for the control of delayed self-regulating and integrating systems based on Laguerre network identification and predictive control. The plant model is arranged in a form that is linear in some parameters that are the weights of each individual Laguerre orthonormal term. Thus, a simple recursive least squares identification scheme suffices to determine the model. The identified model is then used to design a predictive controller. The motivation for this approach was the development of a control tool that is simple and practical for use in an industrial setting. Industrial experience shows that the expected control improvements are in the range of 30 to 50% when compared with PID controller performance for typical deadtime-dominant processes.

The chapter's content is organized in six sections followed by conclusions. The first three sections following this introduction summarize the theory in order to give the reader confidence with the approach taken and also to provide a quick reference document when starting to use the controller. The three sections before conclusions give a walking tour through simulation and a number of industrial examples of the practicalities associated with the implementation of the BrainWave controller. Issues such as model identification in closed-loop or noisy systems are not neglected. Specific references to the modeling and control of integrating type systems are also addressed.

The first section describes in detail the modeling method based on Laguerre functions emphasizing its advantage when compared with other modeling methods. In the second section the formulas used in building adaptive controllers are revealed together with a discussion addressing the simplifications made to ensure the real time implementation of up to 32 multi-input single-output loops simultaneously with a sampling time as low as 0.1 sec. The enhancement of the controller for integrating type processes or unmeasured disturbances with a given characteristic is described in the third section.

The simulation and the field application results presented in the fifth and sixth sections, respectively, are completing the image of the BrainWave controller when deployed in an industrial setting. These sections target the reader interested in the simplicity of implementation and the benefits provided by this controller.

THE LAGUERRE MODELING METHOD

Most adaptive control methods methods use a black box model to determine the relationship between the control variable (CV) and the process variable (PV). Such methods do not use any specific information about the detailed chemical or physical process occurring in the system. The main reason for this approach to process control is the difficulty of obtaining an analytical model due to the complexity of the reactions taking place in the process and its nonlinearity. Also, there are particular processes where the quality of the raw material is determined by outside factors. In such systems the feed is not heterogeneous from batch to batch, resulting in a significant number of unknowns. Therefore such situations are appealing to black box modeling techniques used to determine the plant dynamic model online.

A limitation of these methods can be that the model obtained is linear whereas most processes exhibit a nonlinear characteristic. Still, according to various literature results the black-box method is successful as long as the linear model is valid in the neighborhood of every point on the plant trajectory. In the case of large and rapid variations in the process parameters, such methods might face an implementation barrier. The danger in such situations is that significant DC errors, oscillation, or even instability can occur.

The modeling method used within BrainWave uses discrete-time Laguerre functions to determine at each time step, in a recursive manner, such a linear model of the real plant. The online Laguerre-based identification algorithm has a number of free parameters that are preadjusted for the designer to minimize its efforts. To determine each individual Laguerre orthonormal parameter term a simple recursive least squares (RLS) estimation can be used to solve the optimization problem of choosing the best fit of the Laguerre network that matches the process dynamic response.

Why the Laguerre Method Is Used for Identification

Continuous Laguerre functions have a history of engineering applications of almost 50 years (Lee, 1960; Head, 1956). The motivation for using Laguerre functions as a basis is their simple Laplace representation. Most of the industrial applications use a discrete-time model of the plant. Also, their orthogonality is a major advantage. A reliable solution for the discretization of the continuous-time Laguerre model has been achieved by Zervos (1988).

An essential issue is that the Laguerre state space that models the plant in a linear fashion can reflect only a self-regulating dynamic system response (i.e., stable). In its original form this approach was capable of estimating only self-regulating systems. Hence a further derivation of this algorithm was required and provided to cope with linear time invariant integrating systems.

The main practical advantage of this methodology is the minimal amount of prior knowledge required to commission a loop, essentially a rough estimate of the time delay and the dominant time constant. As seen later, this greatly reduces the time for setup and commissioning. Furthermore, because the time delay is implicitly described by the Laguerre network representation, it is easy to track variations of this parameter. When used in a multimodel system, a Laguerre network with a fixed pole greatly smooths the transition between models, as the state is the same for all models and thus does not require initialization when switching, as long as the models have all the same dimension.

Several theoretical advantages result from the use of an orthonormal series representation of process dynamics, particularly in an adaptive control framework. The Laguerre model is an output-error structure, is linear in the parameters, and preserves convexity for the identification problem, allowing use of a simple recursive least-squares algorithm. The use of an orthonormal series representation effectively eliminates or greatly reduces the parameter drift due to the influence of unmodeled dynamics on the nominal model. The nominal model complexity can easily be changed online with minimal disruption if need be, a very difficult thing to do in the case of a transfer function model. The Laguerre model is stable and is robustly stabilizable as long as the unmodeled dynamics are stable. This effectively solves the so-called admissibility problem (i.e., transfer function models that have unstable pole-zero cancellations) and makes the Laguerre structure particularly suitable for adaptive control applications in the process industries.

For a complete picture we need to say that there are some drawbacks when using orthonormal series in adaptive control. The first and most obvious one is the loss of physical insight. Indeed, poles and, within some limits, zeros are usually easy to interpret and certainly are well known to control engineers. However, it is difficult to give a Laguerre spectrum such immediate physical interpretation. This issue has been partially addressed through the introduction of a model viewer that gives a step response image of the Laguerre network.

Another and more subtle problem comes from the use of an unstructured model and arises when the frequency content of the excitation used during identification is incompatible with the choice of the Laguerre pole and the process dynamics. This can result in artifacts in the identified response. The choice of the optimal Laguerre time scale is discussed in Fu and Dumont (1993). Choosing the Laguerre pole is equivalent to a choice of the model's time scale. This can be currently modified through an adequate value for the sampling time at which the controller is running.

Other Modeling Techniques, Trade-Offs

In terms of black box modeling a common solution is to use autoregressive moving average (ARMA) models. As nothing is perfect, there are problems experienced with these models, generally located at the level of the estimation accuracy. For instance, if the ARMA model has a smaller order than the plant, the estimation of its parameters depends directly on the input signal characteristic. In some cases the resulting model can be unstable. ARMA modeling is also very sensitive to plant model scaling between inputs and outputs. Finally, it is not a simple task to describe systems with a variable deadtime using an ARMA model for adaptive control. One technique is to extend the numerator of the transfer function with as many terms as necessary to cover the expected range of deadtimes at the risk of identifying a model with pole-zero cancellation, or near pole-zero cancellations that will create major problems in adaptive control.

Formulas Used

Whether a discretized (note that a slightly unusual hold must be used to preserve orthonormality) continuous Laguerre network is used, or a discrete Laguerre network such as that of Figure 3.1 and described by

$$L_i(z) = \frac{\sqrt{(1-a^2)}}{z-a} \left(\frac{1-az}{z-a}\right)^{i-1} \tag{3.1}$$

the plant dynamics are represented by the discrete state space

$$\begin{aligned} l(k+1) &= Al(k) + Bu(k) \\ y(k) &= Cl(k) \end{aligned} \tag{3.2}$$

The system output* $y(k) = Cl(k)$ can be thought of as a weighted sum of the Laguerre filter outputs. Hence the plant model can be arranged in a linear form in

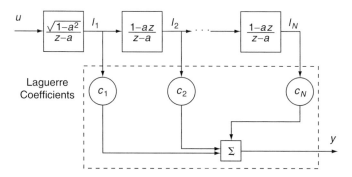

Figure 3.1 *The discrete time Laguerre network of a self-regulating system.*

---

*The model output $y(k)$ represents a small excursion from plant output, plant assumed in equilibrium.

the weights of each individual Laguerre orthonormal term. Recursive least-squares (RLS) can then be used to solve the optimization problem through which the best fit of the Laguerre network for the process dynamic response is found.

Another result used in this work is that any causal and asymptotically stable sampled linear system $G(z)$ can be expressed as

$$G(z) = \sum_{i=1}^{\infty} c_i L_i(z) \qquad (3.3)$$

using a discrete-time Laguerre series, where c_i are called the coefficients of the Laguerre network (see Figure 3.1). Therefore $G(z)$ will be approximated by $\hat{G}(z)$, defined as

$$\hat{G}(z) = \sum_{i=1}^{N} c_i L_i(z) \qquad (3.4)$$

Note that a Laguerre model can be always transformed into a transfer function but a general transfer function can only be approximated by a finite Laguerre orthonormal basis.

The core of the recursive least squares algorithm is the update of the covariance matrix:

$$P(k+1) = \frac{1}{\lambda}\left[P(k) - \frac{P(k)l(k+1)l(k+1)^T P(k)}{\lambda + l(k+1)^T P(k)l(k+1)}\right] + \mu I + \nu P(k)^2 \qquad (3.5)$$

This update includes the additional terms weighted by μ and ν for improved stability of the estimation; see Salgado et al. (1988). Through the introduction of the two additional terms μI and $\nu P(k)^2$ covariance matrix resetting and boundedness are achieved.

Finally the estimate is obtained by adding a correction to the previous estimate. The correction is proportional to the difference between the real output of the plant or disturbance and its prediction based on the previous parameter estimate:

$$C^T(k+1) = C^T(k) + \frac{\alpha P(k)l(k+1)}{\lambda + l(k+1)^T P(k)l(k+1)} e(k) \qquad (3.6)$$

Typical values for these parameters used in this estimator are $\alpha = 0.1$, $\lambda \in [0.9, 0.99]$, $\mu = 0.001$, and $\nu = 0.001$.

Note that for a constant input (i.e., $u(k) = u_{eq}$) the state update will reproduce the previous state and therefore the Laguerre coefficients will be unchanged. This mechanism avoids convergence to wrong values when there is no process-persistent excitation. The algorithm is expected to converge if the model error is small and the input signal "rich" enough.

BUILDING THE ADAPTIVE PREDICTIVE CONTROLLER BASED ON A LAGUERRE STATE SPACE MODEL

Model-based predictive control in our view has a number of appealing attributes: *Simplicity*—the basic idea of MBPC is fairly intuitive and can be understood without advanced mathematics; *Richness*—the common elements of MBPC schemes, such as models, objective functions, and prediction horizons, can be tailored to specific problems; and *Practicality*—the usual combination of linear dynamics and inequality constraints allows realistic nonlinearities to be handled. For these reasons predictive control can remedy some of the drawbacks associated with fixed-gain controllers.

Based on an original theoretical development by Dumont and Zervos (1986; Zervos, 1988; Zervos and Dumont, 1988), the controller was first developed for self-regulating systems. This controller was credited by various users with several features among which we can mention: the reduced effort required to obtain accurate process models, the inclusion of adaptive feedforward compensation, and the ability to cope with severe changes in the process.

These features together with a recognized need in industry created the opportunity for further development of a controller capable of dealing with integrating systems with delay in the presence of unknown output disturbances. These investigations led to an indirect adaptive controller based on identification using an orthonormal series representation working online in conjunction with a predictive controller.

The Concepts behind MBPC

The concept of predictive control involves the repeated optimization of a performance objective (Equation 3.7) over a finite horizon extending from a future time (N_1) up to a prediction horizon (N_2) (Clarke and Mohtadi, 1989; Clarke, 1993).

Figure 3.2 characterizes the way prediction is used within the MBPC control strategy. Given a setpoint $s(k + l)$, a reference $r(k + l)$ is produced by prefiltering and is used within the MBPC cost function:

$$J(k) = \sum_{l=N_1}^{N_2} \|\hat{y}(k+l) - r(k+l)\|_{Q(l)}^2 + \sum_{l=0}^{N_u} \|\Delta u(k+l)\|_{R(l)}^2 \qquad (3.7)$$

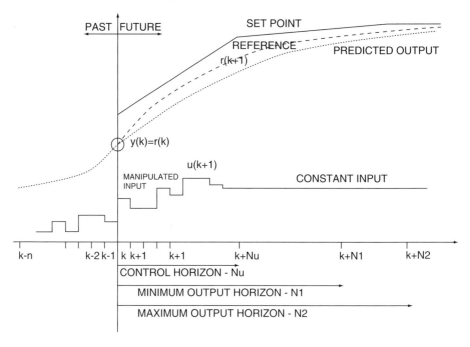

Figure 3.2   The MBPC prediction strategy.

Manipulating the control variable $u(k+l)$ over the control horizon (N_u), the algorithm, as a result of an optimization, over the prediction horizon, drives the predicted output $\hat{y}(k+l)$ toward the reference.

In normal operation the weights $Q(l)$ and $R(l)$ are independent of k. The norm $\|.\|_Q^2$ within the cost function is defined as $\|x\|_Q^2 = x^T Q x$. For prediction it is assumed that $\Delta u(l) = 0$ for $l \geq N_u$. The prediction model is based on the linear frozen representation of the plant model. As formulated, the optimization is a quadratic programming (QP) problem and can be solved using standard algorithms. Further, in the absence of constraints the problem resumes to a simple least squares (LS) problem.

For the BrainWave controller our choice was a simplified version of the predictive control algorithm to ensure a real-time implementation of the whole indirect adaptive scheme, based on a sampling time as low as 0.1 sec for 32 loops simultaneously. As a result input constraints were not managed through an optimization but using a local anti-windup scheme. It is considered (de Dona et al., 1999) that anti-windup has almost similar performance with constrained MBPC for a wide range of plants and control objectives.

The main argument favoring the use of predictive control instead of a conventional state or output feedback control technique is its simplicity in handling varying time delays and nonminimum-phase systems at the same time as input constraints.

A Simple Predictive Control Law

If the coefficients of the discrete Laguerre model in Equation (3.2) are given, we can predict the plant output for the future N_2-steps based on the currently measured output $y(k)$, which is

$$y(k + N_2) = y(k) + C[l(k + N_2) - l(k)] \qquad (3.8)$$

Recursively using the state equation in Equation (3.2) gives the future Laguerre model states:

$$l(k + 1) = Al(k) + Bu(k) \qquad (3.9)$$

$$l(k + 2) = A^2 l(k) + ABu(k) + Bu(k + 1) \qquad (3.10)$$

$$\cdots = \cdots \qquad (3.11)$$

$$l(k + N_2) = A^{N_2} l(k) + A^{N_2-1} Bu(k) + \cdots + Bu(k + N_2 - 1) \qquad (3.12)$$

If we assume

$$u(k) = u(k + 1) = \cdots = u(k + N_2 - 1) \qquad (3.13)$$

(i.e., the future control inputs are held constant), the N_2-step ahead Laguerre model state prediction is then:

$$l(k + N_2) = A^{N_2} l(k) + (A^{N_2-1} + \cdots + I) Bu(k) \qquad (3.14)$$

Substituting (3.14) in Equation (3.8), we finally obtain the d-steps-ahead output prediction:

$$y(k + N_2) = y(k) + C(A^{N_2} - I)l(k) + \beta u(k) \qquad (3.15)$$

where

$$\beta \triangleq C(A^{N_2-1} + \cdots + I)B \qquad (3.16)$$

Setting the N_2-steps-ahead output $y(k + N_2)$ to the desired plant output y_r, we obtain the control $u(k)$ as

Techniques for Adaptive Control

$$u(k) = \frac{y_r - y(k) - C(A^{N_2} - I)l(k)}{\beta} \quad (3.17)$$

It is worth mentioning a few hints for the choice of N_2: (i) unless this horizon is larger than the plant delay, β will be zero; and (ii) if the plant is nonminimum phase, a nonzero β is not guaranteed even when N_2 is larger than the plant delay. Thus N_2 should be sufficiently large. One must choose N_2 such that β is of the same sign as the process static gain and of sufficiently large amplitude.

Therefore a possible criterion to be satisfied when choosing the horizon N_2 is

$$\beta(k)\text{sign}(C(I - A)^{-1}B) \geq \epsilon |C(I - A)^{-1}B| \quad (3.18)$$

with $\epsilon = 0.5$. Note that the matrix $(I - A)^{-1}B$ can be computed offline as it depends only on the Laguerre filters. In adaptive mode, additional computation has to be carried online since the identified model (i.e., the Laguerre coefficient vector C) is changing.

Sometimes, the control signal is expressed in velocity form. Equation (3.17) can be then rewritten as

$$u(k) = (y_r - y(k))/\beta - CS(Al(k) - Al(k-1) - Bu(k-1))/\beta \quad (3.19)$$

Using the definition of β and rearranging, one gets

$$\Delta u(k) = \frac{y_r - y(k) - f^T \Delta l(k)}{\beta} \quad (3.20)$$

where $S = (A^{N_2-1} + \cdots + I)$, $f^T = C^T SA$, $\Delta u(k) = u(k) - u(k-1)$, and $\Delta l(k) = l(k) - l(k-1)$.

The Indirect Adaptive Predictive Control Solution

Indirect adaptive control has certain advantages, when compared with other classic or even robust control techniques:

- Provides online corrections of the model function of the changes in the plant dynamics
- Reduces system sensitivity
- Exhibits simplicity in structure and design
- Deals well with systems with unknown parameters
- Offers an attractive solution for automatic tuning of process control loops

In model-based control the plant model has to be identified in order to produce a control action. Considering the discrete-time Laguerre model we observe that the weights c_i of each individual Laguerre orthonormal term arranged in the row vector C, for a given pole and number of filters, can be selected to approximate the plant or disturbance model.

Here, the proposed indirect adaptive control scheme uses the modified recursive least square (RLS) algorithm Equation (3.6), Equation (3.7) to estimate C. If needed, optimization can be used to find an optimal the Laguerre pole. This is beyond the scope of this presentation.

When we approximate a nonlinear system by a linear model and when the operating condition changes, the approximated model parameters also change. Of course, to correctly estimate the model parameters we work under the assumption that their rate of change is slower than the sampling time T. In this sense a forgetting factor (i.e., an exponentially decaying weight) is added to the measured data sets. The underlying concept for the forgetting factor mechanism is that heavy weighting is assigned to the most recent data because of its importance, versus a small weight in the case of older data.

The properties of the least squares algorithm (i.e., the nonbiased estimation, when there is no model structure error and the measurement noise is white) are readily transferred to the recursive algorithm; see Goodwin and Sin (1984). Also, in the case of no correlation between the output measurement noise and the input sequence, the bias is zero. Enough excitation, even in the presence of feedback, can ensure fulfillment of this condition. When using this algorithm it is mandatory to ensure a "rich" enough input signal. In Åström and Wittenmark (1995) a condition for a signal to be sufficiently rich is stated.

Industrial reality has generated the requirement for an indirect adaptive predictive controller that has also the capability of switching the models used for control and hence the controllers. The control law is computed at each time instant, possibly generating issues of stability and convergence. In Zervos and Dumont (1988) these issues are partially addressed.

Feedforward Variables Used in Modeling and Control

It is easy in the preceding framework to introduce variables to be used to add feedforward compensation to the feedback control law. Feedforward is attractive when a measured disturbance enters the process early in the process. In fact, perfect feedforward compensation is theoretically possible if the delay from the measured disturbance to the plant output is at least equal to that from the manipulated variable to the plant output. Inclusion of a feedforward variable u_f requires an additional Laguerre network:

$$l_f(k+1) = A_f l_f(k) + B_f u_f(k) \tag{3.21}$$

The plant output is then described by

$$y(k) = C^T l(k) + C_f^T l_f(k) \tag{3.22}$$

Following the same derivations as before but using this modified model will give a control law that automatically incorporates feedforward compensation.

Why Simpler Is Better

As mentioned before, a full-fledged model-based predictive control law based on the performance index Equation (3.7) could be used. However, in practice the simple predictive control just derived provides satisfactory performance in most practical situations while being computationally cheap and being easy to use and understand, as it has only one tuning knob, the horizon N_2. The situations where the full-fledged MBPC clearly outperforms the simple one tend to be pathological and are not frequently encountered.

A LAGUERRE-BASED CONTROLLER FOR INTEGRATING SYSTEMS

Time delay integrating systems are common in the process industries. As seen earlier, Laguerre functions are particularly appealing to describe stable dynamic systems. However, they cannot be used to represent dynamics that contain an unstable or an integrating mode. In case of an integrator, we can generally assume knowledge of its presence. We can then simply remove the effect of the integrator from the data by differentiating it. The Laguerre network is then employed to model only the stable part of the plant as in

$$l(k) = Al(k) + Bu(k) \tag{3.23}$$

$$\Delta y(k) = Cl(k) \tag{3.24}$$

where $\Delta y(k) = y(k) - y(k-1)$. Note that (i) $\Delta y(k)$ is an increment and not a deviation from equilibrium for the output, and (ii) the input used in identification has the steady-state value removed (i.e., $u_{true}(k) - u_{eq}$).

When dealing with integrating systems, much more care has to be taken in describing the disturbance affecting the plant and in designing the controller. Marginally stable systems that present time delay have severe inherent bandwidth limitations. Reduced stability margins create a difficulty for classical lead-lag or in the simple case PID controllers. An immediate solution is the use of internal model control which involves the design of a Smith predictor. This solution has its own problems since it requires accurate knowledge of the time delay (a nonrobust solution). The existence of the plant integrator in the loop allows for little error before instability can occur.

The common concept of using a bump test to estimate the process dynamics in the case of an integrating system is not necessarily providing accurate information if the process is not at steady state. Mechanisms to account for these factors have been developed and now are part of the controller, as described in the following.

Identification and Disturbance Estimation for Integrating Systems

To produce an initial model for control, used also in the identification process as starting point, steady-state behavior is requested for the process variable. In the case of an integrating system steady state is achieved only when the contribution of the plant and the disturbance into the process variable (PV) are matching. This is not a common situation; hence a method to remove the integrating characteristic of the response has been developed.

The method employed to estimate the process output slope uses a batch least-squares algorithm. If the slope is smaller than a given threshold the plant is assumed at equilibrium, its input recorded (i.e., u_{eq}), and learning started.

The model states are updated based on

$$l(k+1) = Al(k) + Bu(k) \tag{3.25}$$
$$l_f(k+1) = A_f l_f(k) + B_f u_f(k)$$
$$l_d(k+1) = A_d l_d(k) + B_d u_d(k)$$

and further the output estimations are defined as

$$\hat{y}(k) = \hat{y}(k-1) + C(k)l(k) + \hat{y}_f(k) + \hat{y}_d(k) \tag{3.26}$$
$$\hat{y}_f(k) = \hat{y}_f(k-1) + C_f(k)l_f(k)$$
$$\hat{y}_d(k) = \hat{y}_d(k-1) + C_d(k)l_d(k)$$

where f denotes a feedforward variable and d a variable associated with the unmeasured disturbance model.

Therefore the learning procedure involves an update for the plant and known and unknown disturbance model state estimates Equation (3.25). A flag priority is used to avoid the situation when two models are identified at the same time, leading to a good global prediction but two unusable models for control. Note that the unknown disturbance model can be fixed to a predefined value such as to reflect the types of expected disturbance. Finally, knowing the model states, we can update the plant and known and unknown disturbance model outputs based on Equation (3.26).

Although disturbances can enter the process at any point between the process input and output, they can always be modeled as output disturbances. This is consistent with the Kalman filter approach. Hence, the main practical issue raised is the estimation of the sequence used as an input to the unknown disturbance model. For this an extended Kalman filter uses the $\hat{u}_d(k)$ as in Equation (3.27) and $C_d(k)$ as in Equation (3.6). In practice, at reset, the steps leading to the computation of $\hat{u}_d(k)$ are iterated a number of times for fast convergence.

The main practical issue raised by this approach is how we can estimate the sequence used as an input to the unknown disturbance model $G_d(z)$. Our approach assumes that the unknown $u_d(k)$, unknown since it is unmeasurable, is estimated like

$$u_d(k) = (\hat{y}(k) + \hat{y}_{ff}(k) + \hat{y}_d(k)) - y(k) \qquad (3.27)$$

where $\hat{y}(k)$, $\hat{y}_d(k)$, and $\hat{y}_{ff}(k)$ are the estimated plant, known, and unknown disturbance model outputs. This concept is known in the literature as the extended Kalman filter.

In performing the online model identification the controller checks whether (i) the modeling flag is enabled, (ii) the process variable (PV) $y(k)$ is within the learning range, and (iii) a setpoint (SP) $s(k)$ change in "auto" mode or a control variable (CV) $u(k)$ change in "manual" mode exceeding predefined thresholds has occurred.

The Controller

A simplified version of the MBPC algorithm is used inside the controller to ensure real-time implementation of the whole indirect adaptive scheme, based on a sampling time as low as 0.1 sec. As a result, constraints are not managed through an optimization but using an anti-windup scheme, since it has been shown to have performance almost similar to constrained MBPC (de Dona et al., 1999). The main argument favoring the use of predictive control instead of a conventional state or output feedback control technique is its simplicity in handling varying time delays and nonminimum-phase systems.

The simplified version is characterized by the fact that the N_2 steps ahead output prediction is assumed to have reached the reference trajectory value $r(k + N_2)$. As shown in Figure 3.2 a first-order reference trajectory filter can be employed to define the N_2-steps-ahead setpoint for the predictive controller:

Adaptive Predictive Regulatory Control with BrainWave

$$\hat{r}(k + N_2) = \alpha^{N_2} y(k - 1) + (1 - \alpha^{N_2}) s(k)$$

This condition is not aimed at replacing the equality constraint used in most of predictive control stability proofs, it is just the result of choosing $N_1 = N_2$. The predicted output is equated with the future reference based on

$$r(k + N_2) = \hat{y}(k + N_2 - 1) + \hat{y}_f(k + N_2) + \hat{y}_d(k + N_2) + C(k)l(k + N_2)$$

to obtain the future control move.

Making the assumption that the future command stays unchanged: $u(k) = u(k + 1) = \cdots = u(k + N_2)$ (a condition equivalent to a choice for the control horizon of $N_u = 1$), then the predicted output, equal to the reference N_2 steps ahead, becomes

$$\hat{r}(k + N_2) = y(k - 1) + \hat{y}_f(k - 1) + \hat{y}_d(k - 1) + \\ \delta(k)l(k) + \delta_d(k)l_d(k) + \delta_f(k)l_f(k) + \\ \beta(k)u(k - 1) + \beta_d(k)\hat{u}_d(k) + \beta_f(k)u_f(k) \tag{3.28}$$

where

$$\delta(k) = C(k)A^{N_2}$$
$$\delta_d(k) = C_d(k)A_d^{N_2}$$
$$\delta_f(k) = C_f(k)A_f^{N_2}$$
$$\beta(k) = C(k)(A^{N_2-1} + \cdots + I)B$$
$$\beta_d(k) = C_d(k)(A_d^{N_2-1} + \cdots + I)B_d$$
$$\beta_f(k) = C_f(k)(A_f^{N_2-1} + \cdots + I)B_f$$

Note that here $u(k)$ (the future command) is unknown, $\hat{u}_d(k)$ (the estimated disturbance model input, also known as the feedforward input) is estimated, and $u_f(k)$ (the measured disturbance model input) is measured.

Solving the control equation (3.28) for the future control input $u(k)$ we have:

$$u(k) = \beta(k)^{-1}(r(k + N_2) \\ - (y(k - 1) + \hat{y}_f(k - 1) + \hat{y}_d(k - 1) \\ + \delta(k)l(k) + \delta_d(k)l_d(k) + \delta_f(k)l_f(k) \\ + \beta_d(k)\hat{u}_d(k) + \beta_f(k)u_f(k)))$$

Further, accounting for the internal model principle the plant model needs to be augmented with an integrator for tracking ramps. Ramping signals as references are quite common in the case of batch reactors. Employing the integrator directly in the controller output while setpoint changes are encountered is a common practice:

$$i(k) = i(k-1) + \gamma k_i(r(k) - y(k))$$
$$u(k) = u(k) + i(k)$$

where γ has a nonlinear characteristic to carefully account for a number of updates when the augmented integrator is active following a setpoint change. This strategy has been employed for improved performance because this controller is applied to a wide variety of processes among which only a few have ramping setpoints. In the case of constant setpoints to provide a good disturbance rejection this integrator term is not directly required. The initial state of this integrator is set at u_{eq}. This procedure also takes care of the controller windup when switching between manual and auto modes.

PRACTICAL ISSUES FOR IMPLEMENTING ADAPTIVE PREDICTIVE CONTROLLERS

Integration with Existing Control System Equipment

In most cases, the addition of an adaptive controller to an existing industrial control system requires a new computer to execute the control algorithm, as the existing control system either does not have sufficient computing resources available or is based on a proprietary operating system. In these situations it is unreasonable for users to expect new technologies to be directly integrated into their older control systems. Fortunately, the recent development of standard communications interface protocols for industrial control systems, such as OLE for Process Control (OPC), has improved the ability of users to integrate new technologies on their systems.

The typical concerns that arise when integrating a new computer to an industrial control system are reliability and communication bandwidth. Reliability is an issue because PCs are usually used for data collection or operator interfaces and not involved with real-time process control. Users do not consider PCs to be as reliable as the control system hardware that often features inherent system redundancy. The issue of reliability can be dealt with by installing the adaptive controller in the control system hardware where possible. In addition, the adaptive controller can maintain a heartbeat with the control system to provide a means of automatic fail-over to a backup control scheme or transfer of the adaptive control to a second standby computer.

Communication bandwidth required for the adaptive controller to perform well is related to the process response time (process deadtime plus process time constant). For example, it is not necessary to have process response data received by the adaptive controller every second if the process response time is several hours. For

BrainWave, about 40 data samples in the process response time is adequate for modeling and control. Higher data sample rates may be beneficial in cases where high-frequency noise is present and the data filtering is being performed in the adaptive control computer. Generally anti-aliasing filtering should be done in the control system so the high frequencies are already removed from the data before being sent to the adaptive control computer. In cases where the communication bandwidth is limited, grouping the data packets according to the sample period required to control each particular process can optimize the communication traffic.

Ensuring Successful Identification

The choice of the model order and pole involves a compromise between the accuracy and the cost of attaining it. The plant model accuracy at crossover frequency is very important from the perspective of the closed-loop system transient response. A good choice for the Laguerre pole will be in that frequency region. Further, the choice of the discrete Laguerre function pole a can be restricted to a fixed value, providing a good choice for the system sampling rate T. Choosing an appropriate sampling time, the time scale of the system will be changed. This is the solution adopted in the case of the real-time implementation when for reasons such as speed of the computation, a fixed choice for the pole is required. A database of fixed-pole Laguerre models was built for first-order systems for several choices of deadtime and time constant but fixed pole. This database is used during reset and startup procedures of the commercial controller and reflects a necessary compromise.

In the process industries system identification often has to be performed in closed loop. This is either because the open-loop plant is integrating or even unstable, or simply because of economic considerations running the plant in open loop is not a choice. Of course, in the context of indirect adaptive control one has to estimate the model parameters in closed loop.

Identification in closed loop create difficulties, primarily a possible loss of identifiability due to the fact that the plant input is then strongly correlated with the plant output. There are, however, ways to ensure identifiability in closed loop. The basic principle is to ensure that the plant input does not depend on the plant output via a linear, time-invariant, and noise-free relationship. There are consequently several ways to guarantee identifiability in closed loop:

- Use of a nonlinear controller.
- Switching between different linear controllers. This case is interesting, as it essentially corresponds to the situation in adaptive control when the controller parameters are continuously updated. However, one must note that as those parameters converge to constant values, identifiability slowly disappears. A technique to improve identifiability in predictive control is,

e.g., the implementation of the first two control actions before computing the next set of two, in a pseudo-multirate fashion. In this case, the control law switches continuously between $u(k) = f(r(k), y(k))$ and $u(k) = f(r(k), \hat{y}(k-1)) = g(r(k), y(k-1))$. As shown by Kammer and Dumont (2001) this improves identifiability at minimal cost.
- Use of a known, external excitation to the plant. This is usually realized by way of setpoint changes. A safe procedure is then to only estimate parameters of the plant model following a setpoint change. This can be termed event-triggered identification; see, for instance, the work by Isaksson et al. (2001).

BrainWave uses the setpoint to introduce the excitation. It either waits for a reference change dictated by the operators or, with given acceptance from the customer, is superposing a filtered PRBS signal onto the constant reference.

A benefit of closed-loop identification, however, is that it increases the model precision in the frequency regions that are critical for control design. These frequency regions correspond in time domain to the initial evolution of the plant response after a CV movement has been applied and the plant deadtime has elapsed.

Once the identification procedure has been carried out, the question that arises is, which is the best model for control? Generally the choice is made between the initial model—used to launch the control and identification procedures, respectively—and the newly identified model. An adaptive controller requires a validation algorithm to supervise the identification module and vote on the best option to take. The best plant model is the model that allows the best prediction of the closed-loop behavior around the frequency range where the system operates the most.

Proper data filtering is essential for process model identification in industrial applications. A reasonable choice for a low-pass noise filter would be a filter with a time constant equal to about one-third of the process response time (deadtime plus time constant). This diminishes the detrimental effect of high-frequency noise and unmodeled dynamics.

The most problematic disturbances are those below or near the cutoff frequency of the process. The effects of very low frequency disturbances on process modeling can be reduced by the use of a velocity model and PRBS signal probing (although they are still a problem for identification of integrating processes).

Disturbances near the process cutoff frequency are best handled if there is a measured signal available that correlates to them. In this case, this signal becomes a feedforward in the controller design and the process disturbances caused by changes to this variable become part of the modeled system.

If these disturbances cannot be correlated to a measured signal and all appropriate filtering has been applied, the amplitude of the probing signal must be made as large as possible and data must be collected over a longer period of time. The adaptation rate of the estimator must also be reduced so that the identified model is based on significant amount of data. Generally a signal-to-noise ratio in the range of 5 or better is required to obtain a good Laguerre identification.

Treating Self-Regulating as Integrating Systems

Process transfer functions that can be encountered in industry range from pure gain with zero time constant (such as plug flow control when actuator dynamics are neglected) to pure integrators (such as liquid level control). Between these ideal extremes there exists a continuum of process dynamics with different combinations of deadtime, time constant, and static gain. There are processes that indeed are self-regulating yet have very long time constants and very large static gain. In many of these cases, the steady-state response to even small actuator changes lies beyond the physical boundaries of the system or outside the operating range of interest, so the self-regulating behavior is not observed.

Examples of these systems include heating or cooling of closed systems such as batch chemical reactors. In these cases, the temperature of the heating or cooling media is often several hundred degrees different from the temperature of the reactor contents under control. Because of the closed nature of the system, if more heat is applied to the system than is being lost to the environment, the additional energy will accumulate and the temperature of the system will rise. Because the system is not perfectly insulated, the temperature will rise until the increased rate of energy loss to the environment matches the rate of energy being applied. In practice this temperature may be well beyond the operating range of the reactor. Thus such a system behaves essentially as an open-loop integrator.

From this it is clear that there are many systems that are self-regulating yet in practice can be approximated by integrators. For conventional PID controllers, the range of adjustment of the tuning parameters covers this spectrum in a continuous manner. However, for model-based adaptive predictive controllers, the choice of internal model structure and control equation changes if the controller is to be applied to a system that contains an integrator or not. Self-regulating systems with high static gains and very long time constants can be better controlled if treated as an integrating process. So at exactly what point does one treat a self-regulating system as if it were an integrator?

A simple way to analyze this problem is to consider the desired actuator movement that will result in an appropriate control response. For processes such as level control where the transfer function is just a simple integrator, when a setpoint change is made, the most effective response from the controller would be to either open or close the actuator fully and then immediately return to the equilibrium point. This will be the point where the input flow matches the output flow when the new setpoint is reached. In this ideal case the ratio between the transient move of the actuator and the null steady-state change in actuator position to maintain the new setpoint is infinite. For self-regulating systems, a new steady-state actuator position is required to maintain a new steady-state setpoint. The only reason for making a larger transient move with the actuator is to make the closed-loop response faster than the open-loop one.

Essentially this transient actuator move is dealing with the static accumulation of energy or material required in the system to reach the new operating point, and in this sense this part of the response is similar to an integrating system. In these cases the ratio between the steady-state actuator change and the transient actuator change is typically in the range of 1 to 10.

When that ratio is greater than 10, for model predictive controllers such as Brain-Wave, the system is probably best modeled and controlled as if it were an integrator. Certainly the actuator movements are beginning to approach those desired for a pure integrator as described earlier. In model-based controllers this limit occurs for two practical reasons. First, correctly performing such large transient actuator moves in a predictive control design for self-regulating systems places unreasonable demands on the accuracy of the process model. This is particularly true when the process deadtime is significant compared to the process time constant. Second, because the controller is designed to execute in discrete time, any transient move made by the controller must be held for a minimum of one update interval. Thus the maximum transient move the controller can make is also constrained by the data sample rate, and data sample rate is itself constrained by other design considerations. When a self-regulating system is modeled and controlled as an integrator, the required data sample rate increases significantly as the controller no longer needs to capture and model the long time constant of the self-regulating response. The controller sample rate is now determined just by the process deadtime and the time required for the system to reach an approximately constant rate of change.

Choosing Appropriate Feedforward Variables

New users of advanced controllers have the tendency to want to connect every variable related to a process unit to a controller without analyzing the suitability of these variables as feedforwards. There are several considerations to make when evaluating potential feedforward variables because adding more variables to the control strategy does increase the complexity of the solution (i.e., longer commissioning times and increased maintenance issues). The use of feedforward variables does not come for free since transfer function models must be built, the control strategy must be documented, and often the operating and maintenance practices must be changed to account for the variable, now part of an active controller as opposed to just an indicator.

The feedforward variables must contribute unique information about disturbances on the process. Using more than one variable that is correlated to the same process disturbance not only will complicate the control strategy with no benefit, but also will make the identification of the unique feedforward transfer function models difficult. It is just simpler if only one of the variables is chosen to be a feedforward. This variable will be representative of the disturbance on the process.

In cases where the feedforward variables are changing independently yet have a known relationship to each other, blending the variables into a single calculated feedforward can simplify the control strategy and reduce the process modeling effort required to commission the controller. An example of this situation would be combining a density measurement with a flow rate measurement to produce a single mass flow signal. This approach also makes sense from a process point of view because often such combined variables are more representative of the fundamental cause of the process disturbance and the direct relationship to the process can be better observed. Model-based controllers are often evaluated by potential users based on the number of inputs the controller can accommodate. For BrainWave, three feedforward inputs have been designed into the software and this has been found to be more than adequate, particularly when these blending techniques are used.

There may be feedforward variables that change only very slowly compared to the response time of the process. There is little control performance benefit that can be gained from including such signals in the control strategy. However, there may be some advantage in including such a variable for modeling purposes. Feedforwards that have very long time delays compared to the process response can be delayed by FIFO buffers such that the feedforward variable appears to change much sooner before the process is disturbed. This technique effectively removes a portion of the delay present in the feedforward model and thus allows the controller state update interval to be better matched to the process transfer function dynamics to improve modeling and feedback control performance.

Determining When the Controller Setup Is Correct and Complete

The BrainWave user interface includes both a process data trender and model viewer. These displays can be used together to provide confirmation of correct process model identification and control performance. As setpoint changes are being made during commissioning, the identified process transfer function is plotted in real time on the model viewer. Convergence of the model can be visually confirmed by observing the rate of change of the model as subsequent setpoint changes (process probing) are performed.

With the control prediction horizon set at its high limit (60 control updates ahead), the controller output to setpoint changes is essentially a single step. As the process responds to this actuator change, no further adjustments to the controller output should be required in order to attain setpoint in the future. As the controller tracks the process response and internally compares it to the response predicted by the model, no controller output adjustments should occur if the process and the model predictions match. Simultaneously, the process transfer function model in the model viewer would be observed to be unchanging, also indicating that there was no error in the model predictions requiring changes to the identified model. These simple indications allow the user to confirm that the correct model has been identified and that controller setup is complete. The only remaining adjustment would be to set the controller prediction horizon to a lower value, causing the controller to drive the process to setpoint sooner with more aggressive actuator movements according to the user's preference and physical process considerations.

SIMULATION EXAMPLES

Model Identification in Closed Loop

BrainWave can perform identification of the process transfer function in closed loop. Our preferred method of process excitation is to initiate setpoint changes as opposed to dithering the controller output. Setpoint changes have a more predictable effect on the process because the controller remains in active feedback control during the model identification. The model is computed at each control update of the controller and the newly calculated model is immediately employed for the next control action. The benefit of this approach is that the controller has the opportunity to correct the process model and stabilize the control in cases where the initial process model is substantially incorrect and would cause the controller to become unstable if used.

For BrainWave applications, the first step is to make an estimate of the process dynamics in the form of a first-order plus deadtime (FOPDT) model. This estimate provides a starting point for the controller to validate other settings such as data sample rate and prediction horizon (to ensure that the prediction horizon is at least longer than the process deadtime). The FOPDT estimate is used because of its wide applicability.

In the example shown in Figure 3.3, a FOPDT process estimate has been entered, but the actual process is an underdamped second order process with significant deadtime as shown in Figure 3.4.

Figure 3.5 shows the sequence of setpoint changes and the resulting controller performance as BrainWave adapted the process model under closed-loop control until the model converged to match the actual second-order process. The improving

Figure 3.3   Initial first-order plus deadtime (FOPDT) process model estimate.

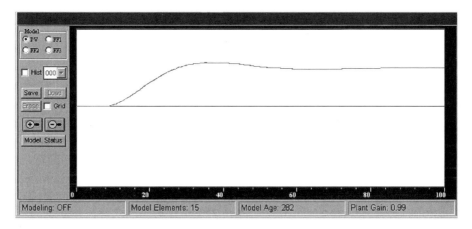

Figure 3.4   Second-order process model identified during closed-loop control.

control performance as the setpoint changes are made also confirms correct model convergence.

Model Identification for Feedforward Variables in Closed Loop

One of the most powerful advantages of BrainWave is the ability to perform identification of the transfer function for feedforward variables during closed-loop control of the process. Feedforward variables provide good excitation for the model identification every time a change occurs, so no additional probing is necessary.

Feedforward control is preferable to feedback control because it provides the opportunity to correct for changes in process conditions before the process deviates from

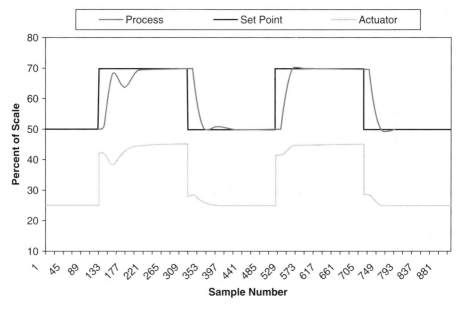

Figure 3.5 Setpoint sequence during closed-loop model identification.

setpoint. Feedback control does not begin to respond until the process is off setpoint when rejecting disturbances. The benefits of feedforward control are also enhanced when dealing with processes with time delay.

In many cases, a given change to the measured feedforward variable does not result in a consistent effect on the process, and the reason for these differences cannot be related to any other observable variable. An example of this would be effluent pH control where the flow rate of the effluent stream is measured as a feedforward variable, but the composition of the stream is unknown and changes under different process conditions. A feedforward control strategy would require that a calculated amount of neutralizing agent be added by the controller for a given change in the effluent flow rate. Because of the changing composition of the effluent stream, any fixed ratio chosen will result in the addition of either too much or too little neutralizing agent. Without another measurement to indicate the composition change in the effluent, the correct feedforward action must be determined based on the process response.

A very effective way to solve feedforward problems of this type is to perform an identification of the feedforward transfer function in a continuous manner while the controller is operating in closed loop. In Figure 3.6, a sequence of step changes in the feedforward are being used by BrainWave to develop a model of how that variable affects the process while operating in closed-loop control. The initial model

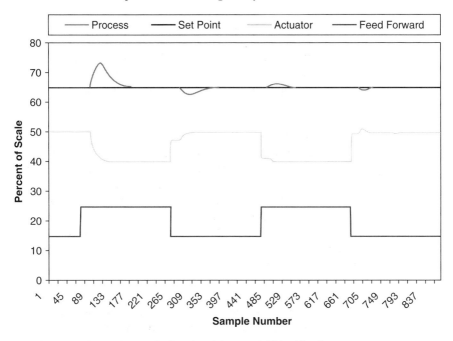

Figure 3.6 Feedforward control using closed-loop model identification.

for the feedforward was chosen to be zero to demonstrate the ability of BrainWave to identify the transfer function from a null model.

The first change in the feedforward variable is rejected entirely by feedback control, as the null feedforward model produced no feedforward control action by the controller. However, the feedforward transfer function model was able to be partially developed based on this data. On the subsequent change to the feedforward variable, the controller uses this model to begin to take some feedforward control action, resulting in a smaller deviation from setpoint. As the feedforward model continues to converge, the feedforward action taken by BrainWave improves, resulting in even smaller deviations from setpoint.

Integrating and Self-Regulating System Control

The control performance of BrainWave can be easily adjusted using simple setup parameters to obtain the desired closed-loop control performance once an accurate process model has been identified. The adjustment of these parameters is made based on various process considerations, such as the ability of the process to accept large transient moves by the actuator and the desired robustness of the controller in the event that the process dynamics change when a fixed process model is used.

124 Techniques for Adaptive Control

For control of self-regulating systems, the primary method to adjust the controller behavior is by the choice of the prediction horizon. This horizon corresponds to the number of controller updates specified for the controller to achieve setpoint. A short prediction horizon results in large transient actuator movements, as these are required to achieve closed-loop time constants that are faster than the open-loop time constant of the process.

The effect of reducing the prediction horizon is shown in Figure 3.7. The first pair of setpoint changes is performed with a prediction horizon of 60 samples, the second pair with a prediction horizon of 40 samples, and the last pair with a prediction horizon of 30 samples. Because of the significant deadtime in the process, the prediction horizon cannot be lowered much further because this will require that the controller make very large transient actuator movements with a long delay before feedback is obtained from the process. This also places unreasonable accuracy requirements on a small area of the process model in the region just beyond the deadtime. At the limit, the prediction horizon cannot be lower than the process deadtime as it becomes physically impossible to achieve setpoint.

For integrating processes, the closed-loop control performance is adjusted in a different manner. By default, BrainWave will attempt to achieve setpoint as quickly as possible, limited only by the allowable range of actuator travel. For these processes, the available performance adjustments are oriented toward reducing

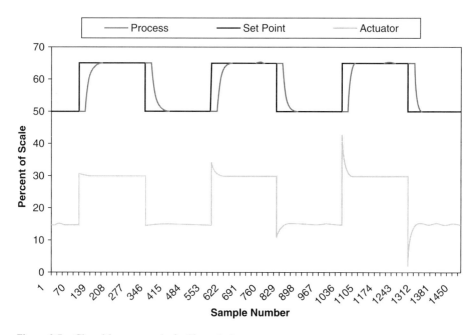

Figure 3.7   Closed-loop control of self-regulating process.

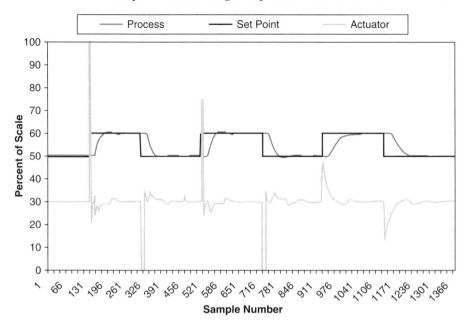

Figure 3.8 Closed-loop control of integrating process.

the control performance from this theoretical ideal. By imposing output limits on the allowable actuator travel, the setpoint change performance can be slowed. Further reductions to the setpoint change performance can be achieved by introducing an internal low-pass filter on the setpoint, in the spirit of a two-degrees-of-freedom controller.

Figure 3.8 shows an integrating process under closed-loop control with BrainWave. The first pair of setpoint changes shows the control performance when the actuator is unrestricted. Note that the complete transient actuator movement is being accomplished well before the process even begins to respond and the full range of actuator travel is being used to achieve setpoint as rapidly as possible. The second pair of setpoint changes shows the controller performance when a high limit of 75% is imposed on the actuator, resulting in a longer elapsed time to achieve setpoint. The third pair of setpoint changes shows the controller performance when a low-pass filter has been applied internally to the setpoint.

INDUSTRIAL APPLICATION EXAMPLES

These examples illustrate a few of the practical problems that industry experiences with the control of processes that exhibit time delay and potentially integrating characteristic using conventional PID controllers. The model-based predictive controller described is providing a practical alternative to PID control. This enables industry to better deal with the common problem of time delay in process control.

Applications to Batch Reactors

Batch production processes are playing an essential role in the chemical and food industry. Many pharmaceutical or biochemical substances and a large number of polymers are produced in such reactors. For a variety of reasons, each batch run is different.

Batch processes that involve heating and cooling exhibit long deadtimes and time constants and have an integrating response due to the circulation of the heating or cooling medium through coils within the reactor, jackets, or a heat exchanger on the outside of the reactor. The reduced stability margin's created by this setup represent a problem for conventional PID controllers. A number of industrial applications of advanced control methods have been reported for such processes. For the most part, those schemes lack the generality required to solve batch reactor industrial control in a unified manner.

The advanced controller described in this chapter is able to model and control such marginally stable processes while accounting for the effect of known and unknown disturbances. The field application results demonstrate that reactors that could previously only be operated manually can now be easily automated, resulting in tremendous improvements in batch consistency, reduced batch cycle times, and therefore increased productivity.

Ethoxylated Fatty Acid Reactor Application

This process involves reacting fat (typically beef tallow) with ethylene oxide (E.O.) to produce a white, tasteless, edible product that is used as a component in many common food products. The reaction is highly exothermic. Ethylene oxide is charged into the E.O. tank at a pressure of about 60 psi and the E.O. feed control valve is opened such that the reactor pressure reaches about 40 psi. The contents of the reactor are circulated through a heat exchanger that can be heated with steam or cooled with water. Initially the reactor contents must be heated above about 140°C before the reaction with the ethylene oxide will start. As the reaction commences, the steam supply is switched out from the heat exchanger and cooling water is switched in. The cooling water flow is then regulated to maintain reactor temperature with a setpoint of about 155°C. As the reaction occurs, E.O. is consumed and the pressure falls below 40 psi. A pressure controller automatically opens the E.O. feed supply valve to maintain reactor pressure at the pressure setpoint. The reaction continues until all of the E.O. is consumed.

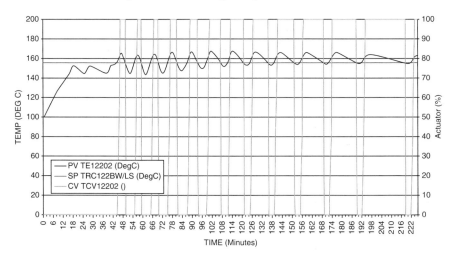

Figure 3.9 Existing PID temperature control performance.

The existing PID-based temperature control was supplemented with override logic in the DCS to prevent high or low temperature excursions. The plant found tuning the PID controller difficult and had to rely extensively upon the override logic to achieve the best temperature control. The temperature of the reactor essentially oscillated between the temperature limits set in the override logic. Despite the obvious oscillation, this scheme produced the smallest range of temperature deviation from the setpoint that could be achieved using a control strategy based on PID. A chart of the existing reactor temperature control is shown in Figure 3.9.

The reaction rate increases significantly as the reactor temperature rises from 155°C to 170°C. Above 170°C, the product begins to discolor, also, the amount of available cooling becomes insufficient. The temperature excursions with the existing control scheme were about 20°C.

The adaptive controller was installed to replace the existing PID temperature controller on the cooling water circuit. The existing DCS temperature control override logic was retained as a safety precaution. Examination of the cooling system utilization under PID control indicated that there was unused cooling capacity available. This was deduced from the duty cycle of the cooling water valve averaging a maximum of about 80% open. Based on this information, the E.O. feed rate pressure control scheme was reassessed. Fundamentally, there was no reason for the reactor pressure to be maintained at 40 psi as the reactor vessel is rated 150 psi, particularly considering that the maximum pressure that could be achieved is equal to the E.O. tank feed pressure of 60 psi. The E.O. feed rate control was thus changed from pressure control to reactor temperature control.

128   **Techniques for Adaptive Control**

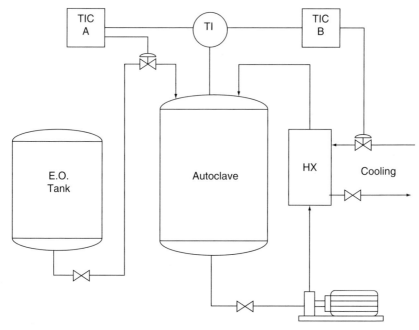

Figure 3.10   Reactor temperature control schematic.

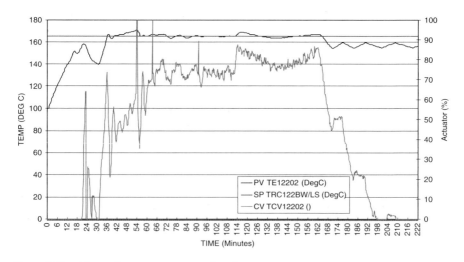

Figure 3.11   Advanced temperature control performance.

A schematic of the new control scheme is shown in Figure 3.10. The complete control strategy consists of the cooling circuit that maintains reactor temperature at 165°C (TIC-B), and the E.O. feed rate control that is simultaneously trying to maintain the reactor temperature at about 170°C (TIC-A). These competing objectives automatically maximize the E.O. feed rate, and thus the production rate of the reactor, as the E.O. feed valve will remain 100% open as long as the cooling system is able to maintain reactor temperature below the high limit of 170°C. If the cooling system is overloaded and the reactor temperature begins to exceed 170°C, the E.O. feed rate is automatically reduced by TIC-A. The adaptive controller was used for both TIC-A and TIC-B. The control loop dynamics exhibited an integrating response with a deadtime of about 800 sec and a time constant of about 1600 sec.

The adaptive controller reduced temperature variability by 75% and was able to maintain reactor temperature within about 5°C of setpoint without excessive operation of the temperature override logic. A chart of the temperature control with the adaptive controller is shown in Figure 3.11.

It is apparent that there is still some cooling capacity available as the cooling water valve experienced a maximum opening of 85%. The improved temperature control allowed the temperature setpoint to be raised from 155°C to 165°C without exceeding the high temperature constraint of 170°C. As mentioned earlier, the reaction rate is highly sensitive to temperature, so the increased average temperature allowed the reaction to be completed much sooner than before.

Figure 3.12 shows a comparison of E.O. feed rates for a batch using the old PID control scheme and the new adaptive control scheme. The E.O. flow for the adaptive

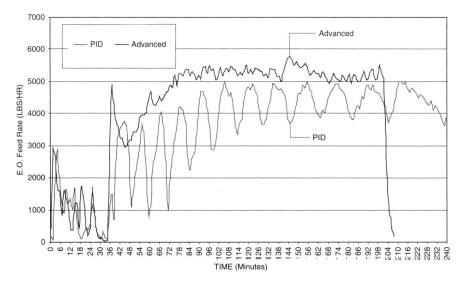

Figure 3.12   Ethylene oxide feed rate comparison.

controller batch maintained consistent feedrates of over 5000 pounds per hour for most of the batch. As a result, the batch was completed about 20% sooner than with the existing PID control scheme, representing a 20% increase in plant capacity. During this batch, the temperature did not exceed the temperature setpoint of 170°C for TIC-A so the E.O. feed control valve remained 100% open for the entire batch. In fact, this indicates that the E.O. tank charge pressure should be raised above 60 psi or the E.O. flow control valve size should be increased to achieve even higher reaction rates to exploit the available cooling system capacity.

DowTherm Batch Reactor Process Control with BrainWave

The chemical batch reactor in this application is used to produce various polyester compounds by combining reagents and then applying heat to the mixture in order to control the reactions and resulting products. A specific temperature profile sequence must be followed to ensure that the exothermic reactions occur in a controlled fashion and that the resulting products have consistent properties. In addition, the reaction rates must be controlled to limit the production of waste gases that must be

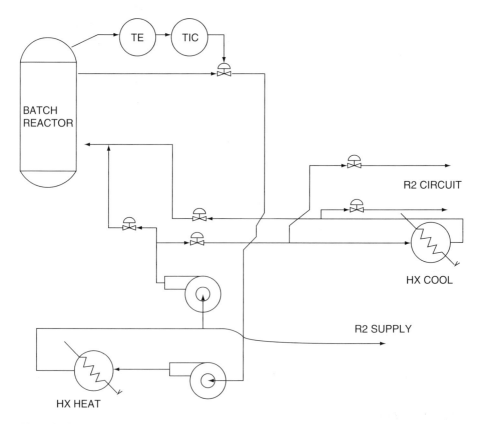

Figure 3.13 The simplified scheme of a batch reactor system.

Adaptive Predictive Regulatory Control with BrainWave   131

incinerated to the design capacity of the incinerator. Because of the potential of an uncontrolled exothermic reaction, proper temperature control is critical to prevent explosions.

The reactor operates in a temperature range between 70 and 220°F and is heated by circulating a fluid (DowTherm-G) through coils on the outside of the reactor. This fluid is itself heated by a natural gas burner to a temperature in the range of 500°F. The reactor temperature control loop monitors temperature inside the reactor and manipulates the flow of the DowTherm fluid to the reactor jacket. It is also possible to cool the reactor by closing the valves on the heat circuit and by recirculating the DowTherm fluid through a second heat exchanger. Cooling is normally only done when the batch is complete to facilitate product handling. A simplified schematic of the system is given in Figure 3.13. The responses of temperatures of the reactor and of the DowTherm fluid at the outlet of the reactor jacket coils are shown in Figure 3.14.

Attempts to automate control of the reactor temperature using a conventional PID controller had been unsuccessful. The reactor temperature is difficult to control because of the long deadtime (about 8 minutes) and long time constant (about 18 minutes) associated with heating the reactor from the outside. This is further complicated because the system essentially behaves as an integrator because of the

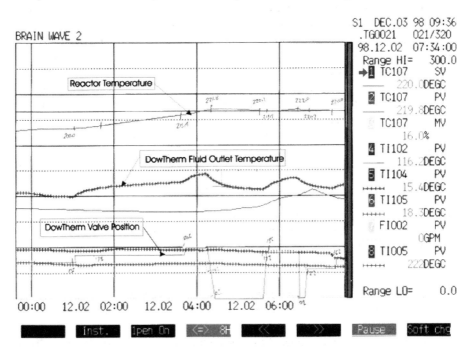

Figure 3.14   The batch reactor system response under manual control.

accumulation of heat in the reactor and is therefore only marginally stable in closed loop. The reactor could only be controlled manually by experienced operators, requiring constant attention to ensure that the temperature profile and resulting reaction rates are correct. It should be noted that a second reactor at this plant was successfully automated using PID control, but this reactor is heated from internal coils and has a much shorter deadtime and time constant and is thus easier to control.

The reactor temperature is stable only if the heat input to the reactor equals the heat losses. If the DowTherm flow is set even slightly higher than this equilibrium point, the reactor temperature will rise at a constant rate until reactor temperature limits are exceeded. The equilibrium point changes during the batch because of heat produced by the exothermic reactions (less energy required to maintain reactor temperature) and the production of vapors (more energy required to maintain reactor temperature). During the final phase of the batch, the exothermic reactions are complete and the vapor production gradually falls almost to zero. Very little energy is required to maintain reactor temperature during this phase.

The operators have developed techniques to manage the manual control of this sequence. From experience, the DowTherm flow is initially set to a nominal value (17% to 19%) that will cause a slow rise in the reactor temperature. The rate of rise is not constant because of the changes in heat requirements that occur during the batch. If the rise is so fast as to cause an overload of the vapor incinerator, or so slow as to stall the temperature rise required to follow the batch profile, the operator will intervene and adjust the flow up or down by 2% to 4%. Otherwise the temperature ramp rate that results from the set DowTherm flow is accepted. During the final phase of the batch, the equilibrium point for the system changes from a DowTherm flow of about 15% to almost 0%. The operators manage this phase by setting the flow to either 20% if the reactor temperature is below setpoint or 0% if the reactor temperature is above setpoint, because these settings will guarantee that the reactor temperature will move in the desired direction. This control method results in oscillation of the reactor temperature about the setpoint and requires constant attention by the operator; see Figure 3.14.

In order to reduce the batch cycle time and improve product consistency, the temperature profile control sequence had to be automated. The inability to obtain automatic closed-loop control of the reactor temperature was a barrier to batch sequence automation.

The BrainWave controller was implemented on the reactor temperature control loop. The controller parameters were estimated from the observed system response from a previous batch and an approximate model of the system was developed. There was some concern that a single model of the system may not be valid for the entire batch sequence because the composition and viscosity of the polyester in the reactor change

Adaptive Predictive Regulatory Control with BrainWave 133

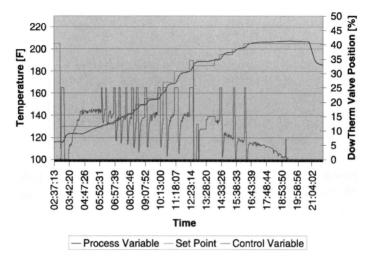

Figure 3.15 The batch reactor system response under automatic control.

substantially during the batch. The first attempt was based on a single model of the reactor response and the control performance was found to be very good. The controller was left in place and has since been controlling the reactor temperature in automatic.

A chart of the temperature control performance of the advanced controller during an entire batch is shown in Figure 3.15. The integrating type response of the reactor is apparent from the control actions made by the controller as the reactor temperature follows the setpoint to higher temperature operating points with a final control output at 0%. Note that the batch sequence was suspended and the controller was placed in manual mode for a short time because of a water supply problem at the plant. The batch sequence was later resumed and the controller was placed in automatic for the rest of the batch.

Operators now adjust the temperature profile setpoint instead of the DowTherm flow. Complete automation of the batch sequence including automatic setpoint ramp generation for the reactor temperature is now possible. Operation of the reactor is also improved because the rate of vapor production is much more constant. This helps to avoid overloading of the vapor incinerator and possible violation of environmental emission regulations due to incomplete combustion of the process waste gases.

The controller was easy to apply and configure. It has achieved very good control performance on a reactor that could not be controlled in a satisfactory manner using PID controls implemented in the plant DCS system. The automatic control of the reactor temperature now enables the plant to reduce batch cycle time, to increase plant productivity, and to improve product quality and consistency through an automated batch sequencer.

Advanced Control of Steam Header Pressure, Saveall Consistency, and Reel Brightness in a TMP Newsprint Mill

Many challenging regulatory control problems can be found in a paper mill. For critical applications, there is a need to achieve increased control performance, but the PID controller is typically the only tool available to provide closed-loop regulatory control. In each case presented, the mill reviewed the tuning parameters of the PID controllers to ensure that the controller was performing at the best level they could achieve. The following examples describe the application of the adaptive controller (AC) to steam header pressure, saveall consistency, and paper reel brightness controls.

Steam Header Pressure Control

The problem with the existing steam header pressure control was the pressure transients that occurred during major upsets to the plant (e.g., the loss of a refiner as major steam source or during a sheet break on the paper machine). The transients would often result in an overpressure condition on the header and venting of steam to atmosphere or a significant underpressure to the mill systems.

A steam accumulator is used to help avoid the underpressure condition as the mill power boiler is unable to respond fast enough to pick up the load when a refiner trips. Refer to Figure 3.16 for a simplified schematic of the pressure control system.

The adaptive controller was installed on the main header pressure control loop (PIC-0460) and the steam accumulator pressure control loop (PIC-0420). The setpoint for PIC-0420 was set slightly lower than PIC-0460 so that the accumulator would only provide steam when the main header pressure could not be maintained. A performance comparison for main steam header pressure control under normal operation is shown in Figure 3.17. The adaptive controller variance from setpoint was about 9 times lower than the one ensured by the PID controller. In addition, the improved control has virtually eliminated the venting of steam during major upsets, resulting is an estimated annual savings of about $50,000.

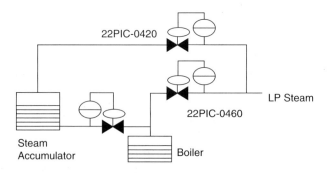

Figure 3.16 Steam header pressure control schematic.

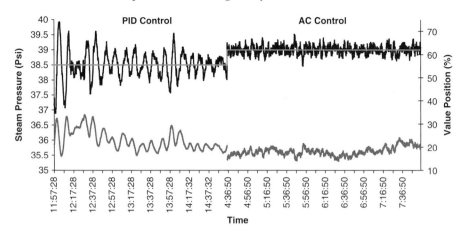

Figure 3.17   Steam header control comparison.

Saveall Consistency Control

The Saveall is a fiber reclaim device that recovers paper fiber from several process streams at the wet end of the paper machine. Saveall consistency control is difficult because of the wide range of variability of feedstock available from the saveall. The flow of white water that is mixed with the stock before it is pumped to the blend chest controls the consistency of the stock recovered from the saveall. A simplified process schematic is shown in Figure 3.18.

During normal operation the dilution is added to the stock and mixed in the saveall mix chest. When additional dilution is required a second valve that operates on a split range from the consistency controller provides dilution at the suction of the saveall mix chest pump during major process upsets.

The mill prefers to add the dilution on the feed into the saveall mix chest as the chest provides additional capacity to minimize the effects of consistency disturbances. The response dynamics of this loop include a 300-sec deadtime and a 900-sec time constant. These dynamics made it difficult for the mill to tune the existing PID controller for good disturbance rejection performance.

It is important to maintain stable consistency in the stock being supplied to the blend chest as disturbances in consistency propagate to the paper machine and affect both basis weight control and paper quality. The advanced controller was able to model the response dynamics and provide improved control performance compared to the existing PID control scheme. A chart of consistency control performance is shown in Figure 3.19 for a 5-hour period of operation for each controller. The adaptive controller was able to reduce the consistency variability by over 86% compared to the existing PID controller.

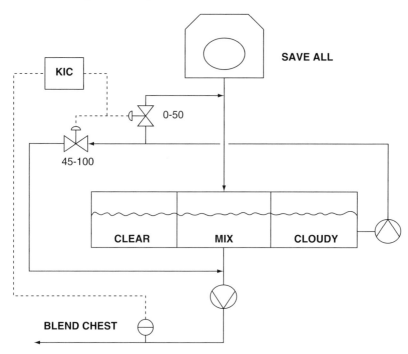

Figure 3.18 Consistency control schematic.

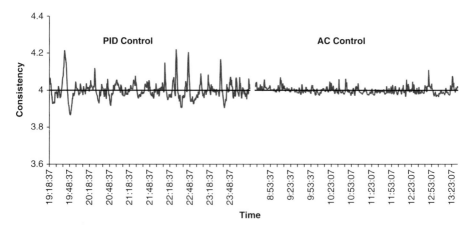

Figure 3.19 Saveall consistency control comparison.

Paper Brightness Control

Paper brightness is an important quality parameter for paper customers. Thermo-mechanical pulp is bleached with sodium hydrosulfite to ensure that the final brightness of the paper at the reel is within the customer's specifications. For many papermakers, the last point in the process where brightness can be corrected

Adaptive Predictive Regulatory Control with BrainWave 137

with bleach is quite far upstream from the reel. For this mill, the response time for a change in bleach to be seen at the reel was about 3 hours. This long response time makes it difficult to achieve closed-loop control based directly on reel brightness. A simplified schematic of the brightness control system is shown in Figure 3.20.

This mill used a manual or "open-loop" control scheme where the operator would monitor brightness at the reel and adjust the setpoint for the compensated brightness controller located at the bleach tube. The operator tends to overbleach to ensure that the reel brightness is above the required minimum, increasing production costs for the paper. The adaptive controller was configured to provide direct, automatic control of reel brightness by adjusting the setpoint of the upstream bleach flow controls. A control performance comparison over a period of about 4 days is shown in Figure 3.21.

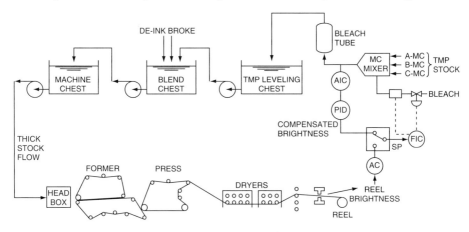

Figure 3.20 Simplified reel brightness control schematic.

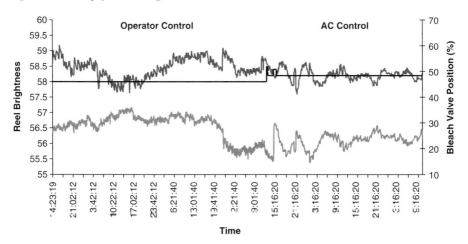

Figure 3.21 Reel brightness control performance comparison.

Results indicate a reduction in reel brightness variability of about 83% compared to the existing control scheme used at the mill. The estimated annual savings in reduced bleach and reduced off-specification paper is about $150,000.

Adaptive Predictive Control of a Glass Forehearth

The production of glass containers presents many control and instrumentation problems. Precise control of the molten glass temperature is particularly difficult because of the changes in the physical properties of the glass as a function of temperature. Typically, the glass temperature must be measured and controlled within 1 to 2°C out of a range of 1100 to 1300°C in order to produce acceptable containers in the molding machine. The molten glass is produced in a large gas-fired furnace from silica sand, soda ash, limestone, and other additives. The furnace uses a combination of surface-fired gas heating and immersed electrode electric heating to melt the raw materials. Because of the poor thermal conductivity of the glass, the furnace operates in an alternating gas fired–heat recuperation cycle to reduce fuel costs. This cycling introduces variations in the temperature of the glass as it exits the furnace and prevents the system from reaching a natural steady state. The molten glass is discharged from the furnace into a distributor where it flows into a number of separate forehearths. Each forehearth has three sections where the glass temperature is measured and controlled.

The section closest to the furnace is called the rear section, followed by the front section, then the conditioning section. The rear section and front section are combined heat/cool zones incorporating natural gas port valves and cooling wind butterfly valves operating inversely to bring the glass to the desired temperature. The last section is the conditioning zone, which is approximately half the length of the two preceding sections and is equipped as a heating zone only (no provision for cooling). The main function of these three zones is to provide a controlled, homogeneous cooling of the glass from the 1500°C of the main tank (furnace) to the production temperature of 1100 to 1170°C. The glass pours out of an orifice in the conditioning section and is sheared into discrete gobs. The gobs are guided as they drop by a series of automated chutes for delivery into the forming machine. A simplified diagram of the furnace and forehearths is shown in Figure 3.22.

The viscosity of the glass is very sensitive to temperature. If viscosity changes, the amount of glass that pours through the gob cutter will change, affecting the resulting weight of the glass container. Container weight is critical for proper molding in the forming machine to obtain the desired container volume and appearance. It is thus a primary quality parameter for the finished container.

For the particular example treated in this section the existing control system used programmable PID controllers that were panel-mounted on the factory floor. A PID

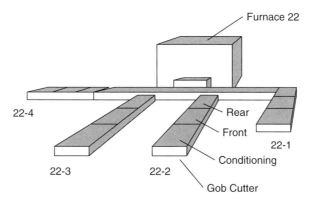

Figure 3.22 Simplified diagram of furnace and forehearths.

controller was used to control the glass temperature in each section of the forehearth. The controllers were also connected to a graphical operator interface via a proprietary network for data acquisition and trending. Control of the molten glass temperature is a difficult problem because (i) the response time of the temperature control loop is quite long (between 20 and 40 minutes), (ii) the combined heating/cooling control actuators are nonlinear, (iii) production rate changes affect the gain and lag time for the temperature control loop, and (iv) thermal and mechanical properties of the glass change with temperature, producing nonlinear dynamics. Operation of the forming machine involves changes in production rate for a given container as well as changes in the type of container to be produced (i.e., a "job change"). Following a production rate change or a job change, the glass temperature would typically take between 4 and 6 hours to settle at the setpoint temperature and enable the forming machine to produce the expected yield of acceptable containers (i.e., "standard pack"). It was not uncommon for the glass temperature to require as long as 10 hours to stabilize. In some cases, the glass temperature continued to oscillate around setpoint for much longer periods and would require several hours of attention by a knowledgeable instrument technician to adjust the PID controller tuning and stabilize the process. Another problem was the operators becoming impatient with the PID controllers and then switching to manual control. This action often prolonged the settling time because the operator would make incorrect control actions.

The adaptive predictive controller was installed at the plant for control of the molten glass temperature on forehearth 22-2. This forehearth was chosen because it was the most difficult to control as a result of its near alignment with the furnace discharge throat. It was exposed to the largest swings in glass temperature from the furnace because the glass spent the least time in the distributor section, which was temperature-controlled and tended to buffer the temperature swings from the furnace. The advanced controller was installed to control glass temperature in all the

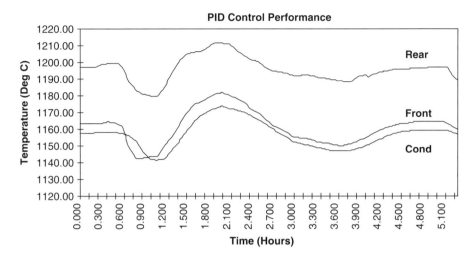

Figure 3.23 PID control performance on a pull change.

sections of forehearth 22-2. The temperature of the glass in the rear section was used as a feedforward for the front section temperature controller. The temperature of the glass in the front section was further used as a feedforward for the conditioning section temperature controller. The feedforward allowed the adaptive controllers to anticipate the control adjustments required to keep the glass temperatures at setpoint as the glass temperature in the preceding sections changed.

The adaptive controller was implemented on a PC platform and was linked to the PID controllers using a serial connection to the existing controller network. Glass temperature, setpoint, and the control mode were read from the PID controllers over the network.

The PID controllers were configured for a "tracking" control mode, which would allow the control actions originating from the adaptive controller to be passed on to the field actuators. This configuration provided the operators with manual, PID, or BrainWave control, allowing the operators to continue to use a familiar interface and providing some security, as the existing control system was still available as a backup to the adaptive controller. When the PID controller was switched to adaptive control, the PID algorithm was bypassed and the adaptive controller assumed control of the process.

Control performance was evaluated based on the time required for the glass temperature to stabilize following a production rate change (i.e., a "pull" change), which typically involves a change in the temperature setpoint. The ability of the adaptive controller to recover following a momentary gas shutoff was also compared to the existing PID controller's performance. The resulting effect of the glass temperature control performance on the yield of acceptable containers pack was then compared.

A plot of the rear, front, and conditioning section temperatures following a pull change with the existing PID controllers is shown in Figure 3.23. The highest temperature is at the rear section and the lowest temperature is at the conditioning section. The temperatures had not settled at setpoint after more than 5 hours following the pull change.

Figure 3.24 shows the performance of the adaptive controller following a similar pull change that also involved a temperature setpoint change. The temperature stabilized in less than 3 hours, which represents about a 50% improvement in temperature settling time compared to PID control. Operating experience has shown that it is not unusual for the PID controllers to take up to 6 hours or more to stabilize temperatures following a pull change. In some cases, the PID controllers would fail to stabilize the temperatures and the controllers would have to be placed in manual mode until they could be retuned by a knowledgeable instrument technician. During this period, the temperature control of the glass would be poor and the yield of acceptable containers would be reduced. By comparison, the adaptive controllers have often been able to stabilize glass temperatures in less than 2 hours and have not required adjustment to maintain performance.

Forehearth 22-2 has about 15 job changes per month. The improved control saves approximately 30 hours/month or 360 hours/year of lost production due to the glass temperature not being stabilized at setpoint. This is about 4% of the annual production of Forehearth 22-2.

The ultimate control performance comparison between the existing PID controllers and the adaptive controllers is their effect on the production of acceptable glass

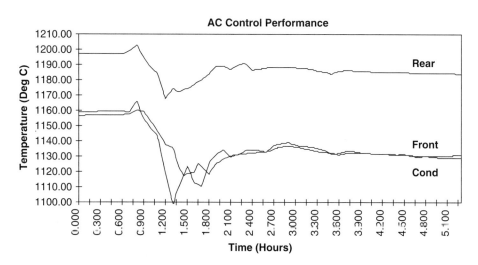

Figure 3.24 Adaptive control (AC) performance on a pull change.

containers. Over the past 2 years of operating experience, the plant has observed an improvement in the standard pack of 3.75% to 20% for the most common containers produced on Forehearth 22–2. Assuming a production rate of 200 containers per minute with a cost of $0.02 per container, this represents a production increase worth $80,000 to $420,000 per year.

CONCLUSION AND LESSONS LEARNED

The main difficulty in designing a predictive controller is to obtain an accurate model and further to tune a multitude of parameters available in the algorithm. We have found that, with experience, systematic tuning procedures based on theoretical results could be developed. Also, the number of parameters involved in the controller can be reduced to a manageable amount. Moreover, using the Laguerre modeling technique, it is possible to obtain satisfactory controller performance even if at startup the predictive controller is based on a very simple model.

When the user's choice is to have the learning off, robust performance for the controller is achieved via detuning the controller's internal model. Further, when particular levels of robust performance are required, in the absence of learning and against large plant perturbations, a multimodel approach is employed with success to ensure satisfactory performance.

The constrained optimization characteristic for most of the predictive controllers is replaced with success by an anti-windup scheme. This approach is reducing the computation time sufficiently to allow real-time constrained operation.

Since progress has been observed in the efficiency of solution algorithms and in the power of the hardware on which they run, we can advocate sample time for the commercial controller as low as 0.1 sec for 32 simultaneous MISO loops.

ACKNOWLEDGMENTS

The work referred in this chapter has been transformed into a commercial product with the exceptional effort of the management and technical staff of Universal Dynamics Technologies Inc., with the leadership and vision of Steve Hagemoen.

REFERENCES

Åström, K. J., and B. Wittenmark (1995). *Adaptive Control*, 2nd ed. Addison Wesley, Reading, MA.
Clarke, D. (1993). *Advances in Model-Based Predictive Control*, pp. 3–21. Oxford University Press.

Clarke, D. W., and C. Mohtadi (1989). "Properties of Generalized Predictive Control." *Automatica*, **25**(6), 859–875.

de Dona, J. A., G. C. Goodwin, and M. M. Seron (1999). "Connections between Model Predictive Control and Anti-Windup Strategies for Dealing with Saturating Actuators." In *Proceedings of the 5th European Control Conference, Karlsruhe, Germany*.

Dumont, G. A., and C. C. Zervos (1986). "Adaptive Controllers based on Orthonormal Series Representation." In *Proceedings of the 2nd IFAC Workshop on Adaptive Systems in Control and Signal Processing, Lund, Sweden*.

Fu, Y., and G. A. Dumont (1993). "An Optimum Time Scale for Discrete Laguerre Network." *IEEE Trans. Automatic Control* **38**, 934–938.

Goodwin, G. C., and K. Sin (1984). *Adaptive Filtering, Prediction and Control*. Prentice-Hall, Englewoods Cliffs, NJ.

Head, J. W. (1956). "Approximation to Transients by Means of Laguerre Series." *Proc. Cambridge Phil. Soc.* **52**, 640–651.

Isaksson, A., A. Horch, and G. Dumont (2001). "Event-Triggered Deadtime Estimation from Closed-Loop Data." In *Proceedings of ACC 2001, Arlington, VA, June 25–27*.

Kammer, L., and G. A. Dumont (2001). "Identification-Oriented Predictive Control." In *Proceedings of the Workshop*. IFAC ALCOSP Workshop, Como, Italy, Aug. 27–29, 2001.

Lee, Y. W. (1960). *Statistical Theory of Communication*, 1st ed. John Wiley & Sons, New York.

Salgado, M. E., G. C. Goodwin, and R. H. Middleton (1988). "Exponential Forgetting and Resetting." *Int. J. Control* **47**(2), 477–485.

Zervos, C. C. (1988). *Adaptive Control Based on Orthonormal Series Representation*. Ph.D. thesis, University of British Columbia.

Zervos, C. C., and G. A. Dumont (1988). "Deterministic Adaptive Control Based on Laguerre Series Representation." *Int. J. Control* **48**(1), 2333–2359.

4

MODEL-FREE ADAPTIVE CONTROL WITH CYBOCON

George S. Cheng

MODEL-FREE ADAPTIVE CONTROL WITH CYBOCON

In this chapter, we will introduce model-free adaptive (MFA) control. The concept of the model-free adaptive control theory and the related issues are discussed. Detailed information regarding the concept, architecture, and algorithms of model-free adaptive (MFA) control is being disclosed and discussed for the first time. The technology is protected by U.S. Patents 6,055,524, 6,360,131, and other pending patents.

CONCEPT OF MFA CONTROL

The formal definition of model-free adaptive (MFA) control is given in this section. A model-free adaptive control system shall be defined to have at least the following properties or features:

- No precise quantitative knowledge of the process is available
- No process identification mechanism or identifier is included in the system
- No controller design for a specific process is needed
- No complicated manual tuning of controller parameters is required
- Closed-loop system stability analysis and criteria are available to guarantee the system stability

The essence of model-free adaptive control can be described with the discussions on the following five issues.

Process Knowledge Issue

Most advanced control techniques for designing control systems are based on a good understanding of the process and its environment. Laplace transfer functions or dynamic differential equations are usually used to represent the process dynamics.

In many process control applications, however, the dynamics may be too complex or the physical process is not well understood. Quantitative knowledge of the process is then not available. This is usually called a "black box" problem.

In many cases, we may have some knowledge of the process but are not sure whether the knowledge is accurate or not. In process control applications, we often deal with raw materials, wild inflows, unpredictable downstream demand changes, and frequent switches of product size, recipe, batch, and loads. These all lead to a common problem: that is, we are not sure if the process knowledge we have is accurate or not. This is usually called a "gray box" problem.

If quantitative knowledge of the process is available, we have a "white box" to deal with. It is a relatively simple task to design a controller for the process in this case because we can use existing, well-established control methods and tools based on the process knowledge.

Although model-free adaptive control can actually deal with black, gray, and white box problems, it is more suitable to deal with the gray box problem, since there is no need to apply a "no-model" control method when a process model is clear, and it is not a good idea to attack a black box problem without making the effort to understand the process.

Process Identification Issue

In most traditional adaptive control methods (model-reference or self-tuning), if the quantitative knowledge of the process is not available, an online or offline identification mechanism is usually required to obtain the process dynamics.

This contributes to a number of fundamental problems such as (i) the headache of offline training that might be required, (ii) the trade-off between the persistent excitation of signals for correct identification and the steady system response for control performance, (iii) the assumption of the process structure, the model convergence, and (iv) system stability issues in real applications.

The main reason why identification-based control methods are not well suited to process control is that control and identification are always a conflict. Good control will lead to a steady state where the key variables setpoint, controller output, and process variable will show straight lines on a trend chart. Since any stable system can reach a steady state, the three straight lines will not have information about the process characteristics. On the other hand, good identification requires persistent excitation of controller output and process variable. That is why identification-based control methods have a tough time being accepted by plant operators, if insertion of noise and disturbances is required to keep the process model updated. Operators just do not like someone else touching their processes unless it is mandatory.

MFA control avoids these fundamental problems by not using any identification mechanism in the system. Once an MFA controller is launched, it will take over control immediately. The algorithms used in MFA controllers to update the weighting factors are based on a sole objective, which is to minimize the error between the setpoint and process variable. That means, when the process is in a steady state where error is close to zero, there is no need to update the MFA controller weighting factors.

Controller Design Issue

One of the main reasons why PID is still so popular in process control is that no complicated controller design procedures are needed to use it. Designing an advanced controller usually requires special expertise and experience. That is one of the main reasons that advanced control methods are not widely used in process control applications.

MFA controllers are developed to be general-purpose controllers. In fact, a series of MFA controllers are designed to control a variety of problematic industrial loops: for instance, SISO MFA to control the processes that PID has difficulty with; nonlinear MFA to control extremely nonlinear processes such as pH or high-pressure loops, anti-delay MFA to control processes with large and varying time delays; MIMO MFA to control multivariable processes; feedforward MFA to deal with large measurable disturbances; and robust MFA to force the process variable to stay within defined bounds.

For a user, there really are no controller design procedures required. One can simply configure the controller with certain parameters and the MFA controller is ready to be launched. This is one of the major differences between a model-free adaptive controller and other model-based advanced controllers.

Controller Parameter Tuning Issue

An adaptive controller should not need to be manually tuned. This is also true of the model-free adaptive controller. However, from an engineering point of view, we have left a tuning parameter open to the user for the purpose of adjusting control performance. Since the MFA controller gain K_c is the only tuning parameter and there is no complex combination of tuning parameters, the user can simply increase the gain to make the controller work more actively and decrease it to detune the controller.

System Stability Issue

Closed-loop control system stability analysis is always an important issue because it determines whether the controller will be practically useful. When the closed-loop system stability criterion is available, one can use the criterion to decide whether the control system can be safely put in operation.

As shown in Figure 4.1, a traditional internal model based adaptive control system has three major components: controller, process, and model.

Here "model" refers to a mathematical representation that can describe the relationship between the process input and output. Self-tuning adaptive control systems typically have this architecture.

The model is usually built by an identification mechanism to minimize the model error E_m, which is the difference between the model output and the process output. The identification mechanism has a learning algorithm to minimize such model error using the process input and output data.

Stability of Model-Based Control System

The stability of the overall closed system is related to the process, the controller, and the identifier in the following manner:

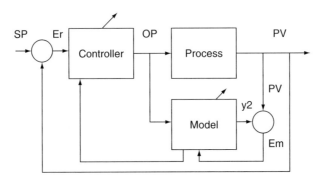

Er=SP-PV Em=y2-PV

Figure 4.1 Internal model-based adaptive control system.

- The stability of the process is usually assumed or we say the process is open-loop stable
- The stability of the control loop must be guaranteed by the convergence of the identifier
- But the convergence of the identifier is dependent on the stability and persistent excitation of signals originating from the control loop

This is a circular argument that it is difficult to resolve. Therefore, many practically useful adaptive controllers are running in a semi-online and semi-offline fashion. That is, although the learning algorithm could be a recursive one that can be implemented in an online fashion, it is not turned on all the time to avoid poor identification results. In a smooth control situation, the system does not provide sufficient excitation signals for good identification.

Since the model-free adaptive controller does not have an identification mechanism, the stability of the overall closed-loop system is much relaxed compared to identification-based control methods. The stability criterion of a MFA control system was derived during the development of model-free adaptive control theory and will be discussed in the following section.

SINGLE-LOOP MFA CONTROL SYSTEM

Figure 4.2 illustrates a single-variable model-free adaptive control system. The structure of the system is as simple as a traditional single-loop control system. It includes a single-input-single-output (SISO) process, an MFA controller, and a feedback loop.

The signals shown in the figure are as follows:

$r(t)$—Setpoint (SP)
$y(t)$—Process variable (PV), $y(t) = x(t) + d(t)$
$x(t)$—Process output, $x(t)$
$u(t)$—Controller output, OP

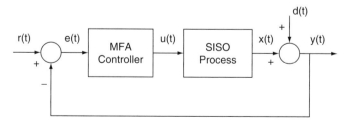

Figure 4.2 Single-loop MFA control system.

$d(t)$—Disturbance caused by noise or load changes
$e(t)$—Error between SP and PV, $e(t) = r(t) - y(t)$

SISO MFA Control Objective

The model-free adaptive controller is an online real-time regulatory controller. Its control objective is to make the process variable $y(t)$ track the given trajectory of its setpoint $r(t)$ under variations of setpoint, disturbance, and process dynamics. In other words, the task of the MFA controller is to minimize the error $e(t)$ in an online fashion.

We select the objective function for MFA control system as

$$E_s(t) = \frac{1}{2}e(t)^2 \\ = \frac{1}{2}[r(t) - y(t)]^2 \qquad (4.1)$$

The minimization of $E_S(t)$ is achieved by (i) the regulatory control capability of the MFA controller, whose output manipulates the manipulated variable forcing the process variable $y(t)$ to track its setpoint $r(t)$; and (ii) the adjustment of the MFA controller weighting factors that allow the controller to deal with the dynamic changes, large disturbances, and other uncertainties of the control system.

SISO MFA Controller Architecture

Figure 4.3 illustrates the architecture of a SISO MFA controller. A multilayer perceptron (MLP) artificial neural network (ANN) is adopted in the design of the controller. The ANN has one input layer, one hidden layer with N neurons, and one output layer with one neuron.

The input signal $e(t)$ to the input layer is converted to a normalized error signal E_1 with a range of -1 to 1 by using the normalization unit, where $N(.)$ denotes a normalization function. The E_1 signal then goes through a series of delay units iteratively, where z^{-1} denotes the unit delay operator. A set of normalized error signals E_2 to E_N is then generated. In this way, a continuous signal $e(t)$ is converted to a series of discrete signals, which are used as the inputs to the ANN. These delayed error signals E_i, $i = 1, \ldots N$, are then conveyed to the hidden layer through the neural network connections. It is equivalent to adding a feedback structure to the neural network. Then the regular static multilayer perceptron becomes a dynamic neural network, which is a key component for the model-free adaptive controller.

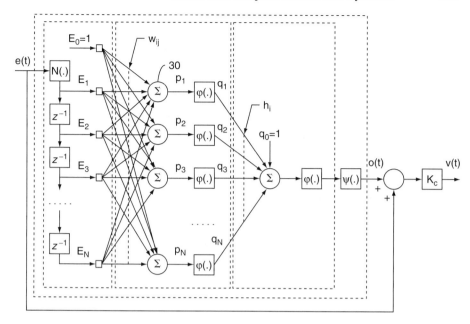

Figure 4.3 Architecture of a SISO MFA controller.

A model-free adaptive controller requires a dynamic block such as a dynamic neural network as its key component. A dynamic block is just another name for a dynamic system, whose inputs and outputs have dynamic relationships.

Each input signal is conveyed separately to each of the neurons in the hidden layer via a path weighted by an individual weighting factor w_{ij}, where $i = 1, 2, \ldots, N$, and $j = 1, 2, \ldots, N$. The inputs to each of the neurons in the hidden layer is summed by adder with $E_0 = 1$, the threshold signal for the hidden layer, through the constant weights $W_{0j} = 1$ to produce signal p_j. Then the signal p_j is filtered by an activation function to produce q_j, where j denotes the jth neuron in the hidden layer.

A sigmoidal function $\varphi(.)$ mapping real numbers to (0,1) defined by

$$\varphi(x) = \frac{1}{1 + e^{-x}} \qquad (4.2)$$

is used as the activation function in the ANN.

Each output signal from the hidden layer is conveyed to the single neuron in the output layer via a path weighted by an individual weighting factor h_i, where $i = 1, 2, \ldots, N$. These signals are summed in adder with $h_0 = 1$, the threshold signal for the output layer, and then filtered by activation function. A function defined by

$$\psi(y) = \ln\frac{y}{1-y} \tag{4.3}$$

maps the range of the output layer from (0,1) back into the real space to produce the output $o(t)$ of the artificial neural network.

SISO MFA Control Algorithm

The algorithm governing the input/output of the controller consists of the following difference equations:

$$p_j(n) = \sum_{i=1}^{N} w_{ij}(n)E_i(n) + 1 \tag{4.4}$$

$$q_j(n) = \varphi(p_j(n)) \tag{4.5}$$

$$o(n) = \psi[\varphi(\sum_{j=1}^{N} h_j(n)q_j(n) + 1)]$$
$$= \sum_{j=1}^{N} h_j(n)q_j(n) + 1 \tag{4.6}$$

$$v(t) = K_c[o(t) + e(t)] \tag{4.7}$$

where n denotes the nth iteration, $o(t)$ is the continuous function of $o(n)$, $v(t)$ is the output of the model-free adaptive controller, and $K_c > 0$, called controller gain, is a constant used to adjust the magnitude of the controller. This constant is useful to fine tune the controller performance or keep the system in stable range.

An online learning algorithm is developed to continuously update the values of the weighting factors of the MFA controller as follows:

$$\Delta w_{ij}(n) = \eta K_c \frac{\partial y(n)}{\partial u(n)} e(n)q_j(n)(1 - q_j(n))E_i(n)\sum_{k=1}^{N} h_k(n) \tag{4.8}$$

$$\Delta h_j(n) = \eta K_c \frac{\partial y(n)}{\partial u(n)} e(n)q_j(n) \tag{4.9}$$

where $\eta > 0$ is the learning rate, and the partial derivative $\partial y(n)/\partial u(n)$ is the gradient of $y(t)$ with respect to $u(t)$, which represents the sensitivity of the output $y(t)$ to variations of the input $u(t)$. It is convenient to define

$$S_f(n) = \frac{\partial y(n)}{\partial u(n)} \qquad (4.10)$$

as the sensitivity function of the process.

Since the process is unknown, the sensitivity function is also unknown. This is the classical "black box" problem that has to be resolved in order to make the algorithm useful.

Through the stability analysis of the model-free adaptive control, it was found that if the process under control is open-loop stable, controllable, and its acting type does not change during the whole period of control, bounding $S_f(n)$ with a set of arbitrary nonzero constants can guarantee the system to be bounded-input-bounded-output (BIBO) stable.

This study implies that the process sensitivity function $S_f(n)$ can be simply replaced by a constant; no special treatment for $S_f(n)$ or any detailed knowledge of the process are required in the learning algorithm of the model-free adaptive controller. By selecting $S_f(n) = 1$, the resulting learning algorithm is as follows:

$$\Delta w_{ij}(n) = \eta K_c e(n) q_j(n)(1 - q_j(n)) E_i(n) \sum_{k=1}^{N} h_k(n) \qquad (4.11)$$

$$\Delta h_j(n) = \eta K_c e(n) q_j(n) \qquad (4.12)$$

Equations (4.1) through (4.12) work for both process direct-acting and reverse acting types. Direct-acting means that an increase in process input will cause its output to increase, and vice versa. Reverse-acting means that an increase in process input will cause its output to decrease, and vice versa. To keep the foregoing equations working for both direct and reverse acting cases, $e(t)$ needs to be calculated differently based on the acting type of the process as follows:

$$e(t) = r(t) - y(t), \text{ if direct acting} \qquad (4.13a)$$

$$e(t) = -[r(t) - y(t)], \text{ if reverse acting} \qquad (4.13b)$$

This is a general treatment for the process acting types. It applies to all model-free adaptive controllers to be introduced later.

SISO MFA Control System Stability Criterion

MFA Control System Stability Criterion

A sufficient condition of MFA control stability can be described as follows: If process is passive, the closed-loop MFA control system stability is guaranteed and the process can be linear, nonlinear, time-invariant, time-varying, single-variable, and multivariable.

This means if the process is open-loop stable, we can guarantee the closed-loop system stability of a SISO MFA control system. Of course, proper setting of its key parameters such as the controller gain K_c is required. In later sections, we will discuss the idea of MFA being a robust controller as well. Since sensitivity and robustness is always a conflict, we designed the MFA controllers to provide adequate sensitivity and robustness. That means MFA is sensitive enough to the variations of its key parameters such as gain K_c. In the mean time, it is robust enough to deal with the uncertainties of process dynamics, load changes, etc., with a much larger robust range than a conventional controller such as PID.

In that sense, MFA can be quite easily placed at its nominal position, and the system stability is guaranteed as long as the uncertainties are within a reasonable range.

Since the criterion is a sufficient condition, for nonpassive processes, we will not know its stability behavior, although the control system might still work. For instance, a nonself-regulating level loop can be easily controlled by an MFA controller. This type of level loop has the integral behavior and is open-loop unstable.

In general, if the process is not open-loop stable, we cannot guarantee the closed-loop system stability. It is always a good idea to stabilize the process first before applying a feedback controller.

Stability Analysis Methods

Two of the most widely used approaches to address stability problems in nonlinear control systems are (i) the stability theory of Lyapunov, and (ii) the input/output stability theory based on functional analysis techniques. Lyapunov stability theory considers stability as an internal property of the system. It basically deals with the effect of momentary perturbations resulting in changes in initial conditions. On the other hand, input/output stability theory, as its name suggests, considers the effect of external inputs to the system.

Lyapunov stability usually deals with a dynamical system described by a nonlinear differential equation in the form of

$$\dot{x}(t) = f(x, t) \tag{4.14}$$

where $x(t_0) = x_0$.

The direct method of Lyapunov enables one to determine whether or not the equilibrium state of a dynamical system in the form in Equation (4.14) is stable without explicitly finding the solution of the nonlinear differential equation. The method has proved effective in the stability analysis of nonlinear differential equations whose solutions are difficult to obtain. It involves finding a suitable scalar function $V(x,t)$ (Lyapunov function) and examining its time derivative $\dot{V}(x,t)$ along the trajectories of the system. The reasoning behind this method is quite simple. In a purely dissipative system, the energy stored in the system is always positive, and time derivative of the energy is nonpositive.

For practical reasons, however, Lyapunov stability analysis methods are not suitable for adaptive systems in many cases, because these methods mainly study the local property of a dynamical system. The definition of Lyapunov stability says that if the state starts sufficiently close to an equilibrium point, the Lyapunov stability will guarantee that the trajectories remain arbitrarily close to the equilibrium point. In adaptive systems, we do not have any control over how close the initial conditions are to the equilibrium values. In addition, in most circumstances, there are no fixed equilibrium points because of the changes in system parameters, etc.

In contrast to the local property of the Lyapunov stability, the input/output stability theory based on functional analysis techniques mainly concerns the input/output properties of nonlinear feedback systems. It is suitable for the stability problems in adaptive systems. The functional analysis approach is also more general. Distributed and lumped systems, discrete and continuous time systems, and single-input-single-output and multi-input-multi-output systems can be treated in a unified fashion.

The concept and important results of input/output stability theory and passivity theory can be found in Desoer and Vidyasagar (1975) and Anderson et al. (1986). We were able to prove that the MFA controller described in this section is passive. Based on the input/output stability theory and passivity theory, we were able to derive a set of theorems as roughly stated in the MFA control system stability criterion. Notice that this is a global stability criterion for a nonlinear control system. Because of the length of this chapter, actual derivations and proofs of this criterion are not included.

MFA Control System Requirements

As a feedback control system, MFA requires the process to have the following behavior:

- The process is controllable
- The process is open-loop stable
- The process is either direct or reverse acting (the process does not change its sign)

156 Techniques for Adaptive Control

If the process is not controllable, one needs to improve the process structure or its variable pairing. This effort might result in a controllable process.

If the process is not open-loop stable, it is always a good practice to stabilize it first. However, for certain simple open-loop unstable processes such as a nonself-regulating level loop, no special treatment is required for MFA to control.

If a process changes its sign within its operating range, it is still possible to design a special controller for this "ill" process. But, it will not be a general-purpose MFA controller.

SISO MFA Configuration

To show how a practical MFA controller is configured, we use a SISO MFA controller configuration screen from CyboCon MFA control software (Figure 4.4). Some of the key fields in this menu are described in the following:

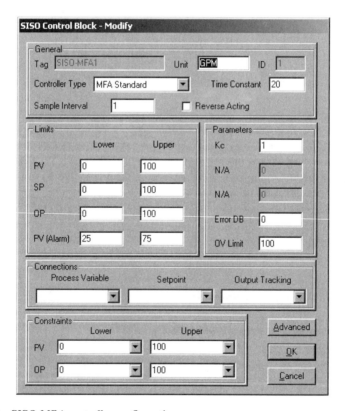

Figure 4.4 SISO MFA controller configuration screen.

- *Time Constant*. A rough estimate of the time constant of the process. Unit: seconds. Range: 0.003 to 99999 sec. Default setting: 20 sec. It is not difficult to estimate the time constant if you know the process.
- *Sample Interval*. The interval between two samples or calculations. Unit: seconds. Range: 0.001 to 999.9 sec. Default setting: 1 sec. According to the principles in the information theory, it is required that the sample interval be less than or equal to one-third of the time constant. That is,

$$T_s \leq (1/3)T_c \qquad (4.15)$$

where T_s is the sample interval, and T_c is the time constant. Notice that the sample interval can be set as small as 0.001 sec, which is 1 ms. CyboCon HS high-speed MFA control software can support this rate for calculation and update of controller output as well as MFA adaptive weighting factors.
- *Reverse Acting*. The process acting type. It is very important to check this field if the process is reverse acting. That means, if the process input increases, the process output decreases, and vice versa.
- K_c. MFA controller gain. Used to adjust the control performance. Set it higher for a more active control action, and set it lower for less overshoot.
- *Advanced Button*. Press this button to enter the menu for configuring feedforward MFA, MFA pH, robust MFA, and nonlinear MFA controllers.

SISO MFA Control Application Guide

MFA Controller Gain

The MFA controller gain K_c is designed to compensate for (i) too large or too small a static gain due to improper scaling; and (ii) severe nonlinear behavior of the process. K_c should be set based on the static gain or the nominal static gain you estimate. The rule of thumb is to make $K_c K \approx 1$.

Based on this general setting, you can adjust the controller gain to fine-tune the control performance—usually, increasing the gain to speed up the control action or decreasing the gain to slow it down. For instance, if you want to see more overshoot, you should let $K_c K > 1$. If you want to see less overshoot, you should let $K_c K < 1$.

Adjusting Time Constant

The time constant in the MFA controller configuration is to provide some qualitative process dynamic information to the controller. It can be just a rough estimate. However, if you feel that the control performance is not as good as expected, you can fine tune the system by adjusting the time constant based on the following rule of thumb.

1. If a controller is reacting to a setpoint change too quickly causing a large overshoot, increase the time constant setting.
2. If a controller is not reacting quickly enough to a setpoint change causing too slow a process response, reduce the time constant setting.

Procedure for Launching MFA Controller

It is easy to start up a new control loop with a SISO MFA controller. Once proper wiring and scaling of setpoint (SP), process variable (PV), and output (OP) are completed, and the estimated process time constant, acting type, and MFA controller gain are entered, the MFA controller can be switched from manual to automatic. The MFA will take control immediately with no bump to the process whether the current process variable is tracking its setpoint or not. In configuring the MFA control system, the controller output should track the current valve position when the controller is in the manual mode. In this way, the MFA is always ready to be switched to the automatic mode without a bump to the process.

MULTIVARIABLE MFA CONTROL SYSTEM

Figure 4.5 illustrates a multivariable feedback control system with a model-free adaptive controller. The system includes a multi-input-multi-output (MIMO) process, a set of controllers, and a set of signal adders, respectively, for each control loop.

The inputs **e(t)** to the controller are presented by comparing the setpoints **r(t)** with the process variables **y(t)**, which are the process responses to controller outputs **u(t)** and the disturbance signals **d(t)**. Since it is a multivariable system, all the signals here are vectors represented in bold type as follows.

$$\mathbf{r}(t) = [r_1(t), r_2(t), \ldots, r_N(t)]^T \quad (4.16a)$$

$$\mathbf{e}(t) = [e_1(t), e_2(t), \ldots, e_N(t)]^T \quad (4.16b)$$

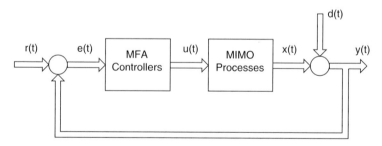

Figure 4.5 Multivariable MFA control system.

$$\mathbf{u}(t) = [u_1(t), u_2(t), \ldots, u_N(t)]^T \qquad (4.16c)$$

$$\mathbf{y}(t) = [y_1(t), y_2(t), \ldots, y_N(t)]^T \qquad (4.16d)$$

$$\mathbf{d}(t) = [d_1(t), d_2(t), \ldots, d_N(t)]^T \qquad (4.16e)$$

where superscript T denotes the transpose of the vector, and subscript N denotes the total element number of the vector.

2-Input-2-Output MFA Control System

Without losing generality, we will show how a multivariable model-free adaptive control system works with a 2-input-2-output (2×2) system as illustrated in Figure 4.6, which is the 2×2 arrangement of Figure 4.5. In the 2×2 MFA control system, the MFA controller set consists of two controllers C_{11}, C_{22}, and two compensators C_{21} and C_{12}. The process has four subprocesses G_{11}, G_{21}, G_{12}, and G_{22}.

The measured process variables y_1 and y_2 are used as the feedback signals of the main control loops. They are compared with the setpoints r_1 and r_2 to produce errors e_1 and e_2. The output of each controller associated with one of the inputs e_1 or e_2 is combined with the output of the compensator associated with the other input to produce control signals u_1 and u_2. The output of each subprocess is cross-added to produce measured process variables y_1 and y_2. Notice that in real applications the outputs from the subprocesses are not measurable and only their combined signals y_1 and y_2 can be measured. Thus, by the nature of the 2×2 process, the inputs u_1 and u_2

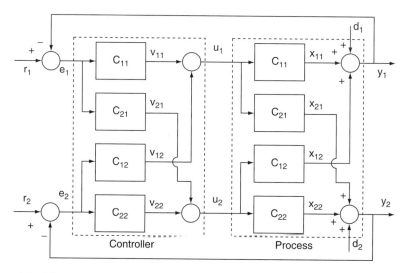

Figure 4.6 2-input-2-output MFA control system.

to the process are interconnected with its outputs y_1 and y_2. The change in one input will cause both outputs to change.

In this 2×2 system, the element number N in Equation (4.16) is equal to 2 and the signals shown in Figure 4.6 are as follows:

$r_1(t), r_2(t)$—Setpoint of controllers C_{11} and C_{22}, respectively
$e_1(t), e_2(t)$—Error between the setpoint and process variable
$v_{11}(t), v_{22}(t)$—Output of controller C_{11} and C_{22}, respectively
$v_{21}(t), v_{12}(t)$—Output of compensators C_{21} and C_{12}, respectively
$u_1(t), u_2(t)$—Inputs to the process, or the outputs of the 2×2 controller set
$x_{11}(t), x_{21}(t), x_{12}(t), x_{22}(t)$—Output of process G_{11}, G_{21}, G_{12}, and G_{22}, respectively
$d_1(t), d_2(t)$—Disturbance to y_1 and y_2, respectively
$y_1(t), y_2(t)$—Process variables of the 2×2 process

The relationship between these signals is as follows:

$$e_1(t) = r_1(t) - y_1(t) \tag{4.17a}$$

$$e_2(t) = r_2(t) - y_2(t) \tag{4.17b}$$

$$y_1(t) = x_{11}(t) + x_{12}(t) \tag{4.17c}$$

$$y_2(t) = x_{21}(t) + x_{22}(t) \tag{4.17d}$$

$$u_1(t) = v_{11}(t) + v_{12}(t) \tag{4.17e}$$

$$u_2(t) = v_{21}(t) + v_{22}(t) \tag{4.17f}$$

2 × 2 MFA Control Objective

The control objectives for this 2×2 MFA control system are to produce control outputs $u_1(t)$ and $u_2(t)$ to manipulate their manipulated variables so that the process variables $y_1(t)$ and $y_2(t)$ will track their setpoints $r_1(t)$ and $r_2(t)$, respectively.

The minimization of $e_1(t)$ and $e_2(t)$ is achieved by (i) the regulatory control capability of the MFA controllers, whose outputs manipulate the manipulated variables, forcing the process variable $y(t)$ to track its setpoints $r_1(t)$ and $r_2(t)$; (ii) the decoupling capability of the MFA compensators, whose outputs are added to the MFA controller outputs to compensate for the interactions from the other subprocess; and (iii) the adjustment of the MFA controller weighting factors that allow the controllers to deal with the dynamic changes, large disturbances, and other uncertainties of the control system.

Since these two loops are interacting to each other, achieving the control objectives is not easy in comparison to a single-loop control system.

2 × 2 MFA Control Algorithm

The controllers C_{11} and C_{22} have the same structure as the SISO MFA controller shown in Figure 4.3. The input and output relationship in these controllers is represented by the following equations:

For controller C_{11}:

$$p_j^{11}(n) = \sum_{i=1}^{N} w_{ij}^{11}(n) E_i^{11}(n) + 1 \tag{4.18}$$

$$q_j^{11}(n) = \varphi(p_j^{11}(n)) \tag{4.19}$$

$$v_{11}(n) = K_c^{11} \left[\sum_{j=1}^{N} h_j^{11}(n) q_j^{11}(n) + 1 + e_1(n) \right] \tag{4.20}$$

$$\Delta w_{ij}^{11}(n) = \eta^{11} K_c^{11} e_1(n) q_j^{11}(n)(1 - q_j^{11}(n)) E_i^{11}(n) \sum_{k=1}^{N} h_k^{11}(n) \tag{4.21}$$

$$\Delta h_j^{11}(n) = \eta^{11} K_c^{11} e_1(n) q_j^{11}(n) \tag{4.22}$$

For controller C_{22}:

$$p_j^{22}(n) = \sum_{i=1}^{N} w_{ij}^{22}(n) E_i^{22}(n) + 1 \tag{4.23}$$

$$q_j^{22}(n) = \varphi(p_j^{22}(n)) \tag{4.24}$$

$$V_{22}(n) = K_c^{22} \left[\sum_{j=1}^{N} h_j^{22}(n) q_j^{22}(n) + 1 + e_2(n) \right] \tag{4.25}$$

$$\Delta w_{ij}^{22}(n) = \eta^{22} K_c^{22} e_2(n) q_j^{22}(n)(1 - q_j^{22}(n)) E_i^{22}(n) \sum_{k=1}^{N} h_k^{22}(n) \tag{4.26}$$

$$\Delta h_j^{22}(n) = \eta^{22} K_c^{22} e_2(n) q_j^{22}(n) \tag{4.27}$$

162 Techniques for Adaptive Control

In these equations, $\eta^{11} > 0$ and $\eta^{22} > 0$ are the learning rate, and $K_c^{11} > 0$ and $K_c^{22} > 0$ are the controller gain for C_{11} and C_{22}, respectively. $E_i^{11}(n)$ is the delayed error signal of $e_1(n)$ and $E_i^{22}(n)$ is the delayed error signal of $e_2(n)$.

The architecture of the compensators C_{12} and C_{21} is shown in Figure 4.7. This architecture differs from the structure of the SISO MFA controller of Figure 4.3 in that no error signal is added to the neural network output $o(t)$.

The input and output relationship in these compensators is represented by the following equations:

For compensator C_{21}:

$$p_j^{21}(n) = \sum_{i=1}^{N} w_{ij}^{21}(n) E_i^{21}(n) + 1 \qquad (4.28)$$

$$q_j^{21}(n) = \varphi(p_j^{21}(n)), \qquad (4.29)$$

$$v_{21}(n) = K_s^{21} K_c^{21} \left[\sum_{j=1}^{N} h_j^{21}(n) q_j^{21}(n) + 1 \right] \qquad (4.30)$$

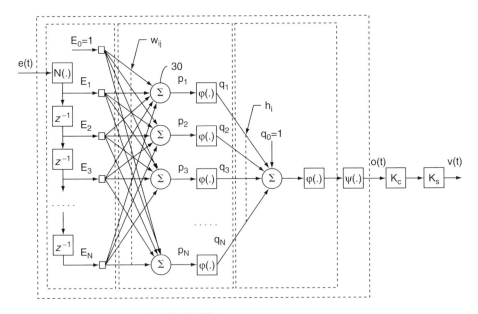

Figure 4.7 *Architecture of a MIMO MFA compensator.*

$$\Delta w_{ij}^{21}(n) = \eta^{21} K_c^{21} e_1(n) q_j^{21}(n)(1 - q_j^{21}(n)) E_i^{21}(n) \sum_{k=1}^{N} h_k^{21}(n) \quad (4.31)$$

$$\Delta h_j^{21}(n) = \eta^{21} K_c^{21} e_1(n) q_j^{21}(n) \quad (4.32)$$

For compensator C12:

$$p_j^{12}(n) = \sum_{i=1}^{N} w_{ij}^{12}(n) E_i^{12}(n) + 1 \quad (4.33)$$

$$q_j^{12}(n) = \varphi(p_j^{12}(n)) \quad (4.34)$$

$$v_{12}(n) = K_s^{12} K_c^{12} \left[\sum_{j=1}^{N} h_j^{12}(n) q_j^{12}(n) + 1 \right] \quad (4.35)$$

$$\Delta w_{ij}^{12}(n) = \eta^{12} K_c^{12} e_2(n) q_j^{12}(n)(1 - q_j^{12}(n)) E_i^{12}(n) \sum_{k=1}^{N} h_k^{12}(n) \quad (4.36)$$

$$\Delta h_j^{12}(n) = \eta^{12} K_c^{12} e_2(n) q_j^{12}(n) \quad (4.37)$$

In these equations, $\eta^{21} > 0$ and $\eta^{12} > 0$ are the learning rate, and $K_c^{21} > 0$ and $K_c^{12} > 0$ are the controller gain, for C_{21} and C_{12} respectively. $E_i^{21}(n)$ is the delayed error signal of $e_1(n)$ and $E_i^{12}(n)$ is the delayed error signal of $e_2(n)$.

The compensator sign factors K_s^{21} and K_s^{12} are a set of constants relating to the acting types of the process as follows:

$$K_s^{21} = 1, \text{ if } G_{22} \text{ and } G_{21} \text{ have different acting types} \quad (4.38a)$$

$$K_s^{21} = -1, \text{ if } G_{22} \text{ and } G_{21} \text{ have the same acting type} \quad (4.38b)$$

$$K_s^{12} = 1, \text{ if } G_{11} \text{ and } G_{12} \text{ have different acting types} \quad (4.38c)$$

$$K_s^{12} = -1, \text{ if } G_{11} \text{ and } G_{12} \text{ have the same acting type} \quad (4.38d)$$

These sign factors are needed to ensure that the MFA compensators produce signals in the correct direction so that the disturbances caused by the coupling factors of the multivariable process can be reduced.

N-Input-N-Output MFA Control System

A 3 × 3 multivariable model-free adaptive control system is illustrated in Figure 4.8 with a signal flow chart.

Techniques for Adaptive Control

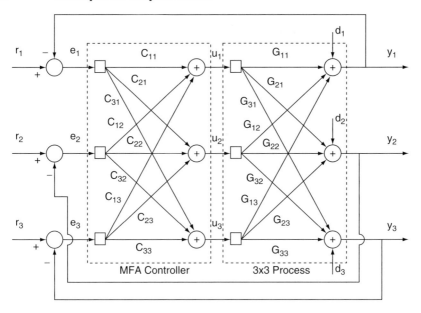

Figure 4.8 3-input-3-output MFA control system.

In the 3 × 3 MFA control system, the MFA controller set consists of three controllers C_{11}, C_{22}, C_{33}, and six compensators C_{21}, C_{31}, C_{12}, C_{32}, C_{13}, and C_{23}. The process has nine subprocesses G_{11} through G_{33}. The process variables y_1, y_2, and y_3 are used as the feedback signals of the main control loops. They are compared with the setpoints r_1, r_2, and r_3 to produce errors e_1, e_2, and e_3. The output of each controller associated with one of the inputs e_1, e_2, or e_3 is combined with the output of the compensators associated with the other two inputs to produce control signals u_1, u_2, and u_3.

Without losing generality, a set of equations that apply to an arbitrary $N \times N$ multivariable model-free adaptive control system is given in the following. If $N = 3$, it applies to the above-stated 3 × 3 MFA control system.

For controller C_{ll}:

$$p_j^{ll}(n) = \sum_{i=1}^{N} w_{ij}^{ll}(n) E_i^{ll}(n) + 1 \qquad (4.39)$$

$$q_j^{ll}(n) = \varphi(p_j^{ll}(n)) \qquad (4.40)$$

$$v_{ll}(n) = K_c^{ll} \left[\sum_{j=1}^{N} h_j^{ll}(n) q_j^{ll}(n) + 1 + e_l(n) \right] \qquad (4.41)$$

$$\Delta w_{ij}^{ll}(n) = \eta^{ll} K_c^{ll} e_l(n) q_j^{ll}(n)(1 - q_j^{ll}(n)) E_i^{ll}(n) \sum_{k=1}^{N} h_k^{ll}(n) \qquad (4.42)$$

$$\Delta h_j^{ll}(n) = \eta^{ll} K_c^{ll} e_l(n) q_j^{ll}(n) \qquad (4.43)$$

where $l = 1, 2, \ldots, N$.

For compensator C_{lm}:

$$p_j^{lm}(n) = \sum_{i=1}^{N} w_{ij}^{lm}(n) E_i^{lm}(n) + 1 \qquad (4.44)$$

$$q_j^{lm}(n) = \varphi(p_j^{lm}(n)) \qquad (4.45)$$

$$v_{lm}(n) = K_s^{lm} K_c^{lm} \left[\sum_{j=1}^{N} h_j^{lm}(n) q_j^{lm}(n) + 1 \right] \qquad (4.46)$$

$$\Delta w_{ij}^{lm}(n) = \eta^{lm} K_c^{lm} e_m(n) q_j^{lm}(n)(1 - q_j^{lm}(n)) E_i^{lm}(n) \sum_{k=1}^{N} h_k^{lm}(n) \qquad (4.47)$$

$$\Delta h_j^{lm}(n) = \eta^{lm} K_c^{lm} e_m(n) q_j^{lm}(n) \qquad (4.48)$$

where $l = 1, 2, \ldots N; m = 1, 2, \ldots, N;$ and $l \ne m$.

In these equations, $\eta^{ll} > 0$ and $\eta^{lm} > 0$ are the learning rate, $K_c^{ll} > 0$ and $K_c^{lm} > 0$ are the controller gain, for C_{ll} and C_{lm}, respectively. $E_i^{ll}(n)$ is the delayed error signal of $e_l(n)$ and $E_i^{lm}(n)$ is the delayed error signal of $e_m(n)$.

K_s^{lm} is the sign factor for the MFA compensator, which is selected based on the acting types of the subprocesses as follows:

$$K_s^{lm} = 1, \text{ if } G_{ll} \text{ and } G_{lm} \text{ have different acting types} \qquad (4.49a)$$

$$K_s^{lm} = -1, \text{ if } G_{ll} \text{ and } G_{lm} \text{ have the same acting type} \qquad (4.49b)$$

where $l = 1, 2, \ldots, N; m = 1, 2, \ldots, N;$ and $l \ne m$.

MIMO MFA Configuration

To show how a practical multivariable MFA controller is configured, we use a MIMO MFA controller configuration screen from CyboCon MFA control software

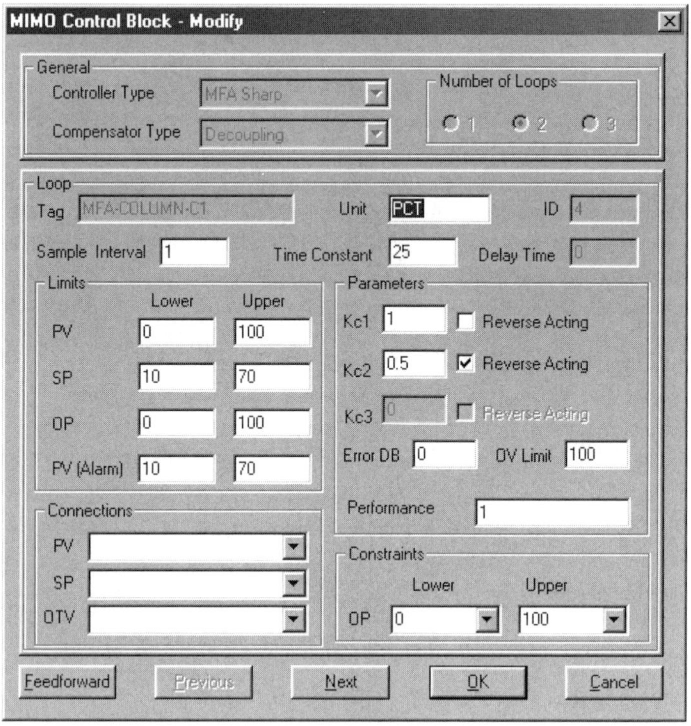

Figure 4.9 MIMO MFA controller configuration screen.

as shown in Figure 4.9. Some key variables in this menu are described in the following:

- *Compensator Type.* Specify the MIMO controller's compensator type. These are the optional controllers: decoupling, anti-delay, and combined (decoupling + anti-delay). The decoupling type is suitable for a 2×2 or 3×3 process without large time delays. If your process has large time delays, you need to specify the anti-delay option for the SISO process, or combined option for the MIMO process.
- *Sample Interval.* The interval between two samples or calculations. Unit: seconds. Range: 0.001 to 999.9 sec. Default setting: 1 sec. According to the principles in the information theory, it is required that the sample interval be smaller than or equal to one-third of the time constant. That is,

$$T_s \leq (1/3)T_c \qquad (4.50)$$

where T_s is the sample interval, and T_c is the time constant.

- *Time Constant.* A rough estimate of the time constant of the main process. Unit: seconds. Range: 0.003 to 99,999 sec. Default setting: 20 sec. For example, if you are configuring C1 of a MIMO controller, this field is related to the time constant of process G_{11}.
- *Delay Time.* This field is not applicable for MIMO decoupling controllers. For the MIMO combined controllers, this field is related to the delay time of the main process. For example, if you are configuring C1 of a MIMO combined controller, this field is related to the delay time of process G11. For the MFA combined controller, this field specifies the process delay time between its input and output actions. Unit: second. If you do not know the delay time, look for the trends of the OP and PV signals. Estimate the delay time between a change in the OP signal and its corresponding response in the PV signal. You may need to adjust the delay time after running the controller to improve performance.
- K_{c1}. MFA controller gain for the main controller. Used to adjust the control performance for the main loop. Set it higher for a more active control action, and set it lower for less overshoot.
- *Reverse Acting* (for K_{c1}). The process acting type in the main loop. It is very important to check this field if the process is reverse acting. That means, if the process input increases, the process output decreases, and vice versa. For example, if you are configuring C1 of a MIMO controller, you need to check this field if G_{11} is reverse acting. If you are configuring C2 of a MIMO controller, you need to check this field if G_{22} is reverse acting.
- K_{c2}, K_{c3}. MFA compensator gain to deal with the interaction from other loops. Used to adjust the decoupling factors. Set it higher for a more active decoupling action, set it lower for less decoupling action, and set it to 0 to disable the compensator.
- *Reverse Acting* (for K_{c2}, K_{c3}). The process acting type. It is very important to check this field if the process is reverse acting. That means, if the process input increases, the process output decreases, and vice versa. For example, for a 2 × 2 system, if you are configuring C1, you need to check this field for K_{c2} if G_{12} is reverse acting.

MIMO MFA Application Guide

A multivariable process has multiple interactive inputs and outputs. For instance, a 2-input-2-output process (or 2 × 2 process) as illustrated in Figure 4.10 has two manipulated variables and two controlled variables. Changing one manipulated variable will affect both controlled variables. This phenomenon is called loop interaction.

Without loss of generality, we will discuss the principles of a multivariable process and related control system using only the case of a 2 × 2 process.

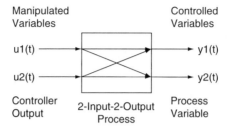

Figure 4.10 2-input-2-output process.

Relative Gain

The Bristol's relative gain is a simple yet powerful measure of the degree of loop interaction. Relative gain is defined as the ratio of open-loop gain (K_{ij}) to the closed-loop gain (K'_{ij}) (Hang et al., 1993).

The open-loop gains K_{ij} are the static gains of process G_{ij}. For instance, K_{11} is the static gain of G_{11}, and so on.

The closed-loop gains K'_{ij} are defined as the ratios of the change in one process output (for instance, y_i), to its input (u_i), when the other process output y_j is kept constant by manipulating u_j.

The relative gain is defined as

$$R_{ij} = \frac{K_{ij}}{K'_{ij}} \qquad (4.51)$$

where K_{ij} is the process open-loop gain, and K'_{ij} is the process closed-loop gain. R_{ij} can be analytically computed by the following formula:

$$R_{11} = R_{22} = R = R_{ij} = \frac{K_{11}K_{22}}{K_{11}K_{22} - K_{12}K_{21}} \qquad (4.52)$$

$$R_{12} = R_{21} = 1 - R \qquad (4.53)$$

By ignoring the process dynamic elements such as the time constant and time delays, relative gains basically reflect the steady-state behavior of a multivariable process.

Pairing Rules of MIMO System

When designing a multivariable control system, the first step is to decide which controlled variable is paired with a manipulated variable.

Pairing Rule 1

Each process of the main loops has to be (i) controllable, (ii) open-loop stable, and (iii) either reverse or direct acting.

Pairing Rule 2

The measure of relative gains can help decide the correct pairing based on process static behavior.

1. A process with a large static gain should be included in the main loop as the main process (for instance, G_{11}, G_{22}).
2. A process with a small static gain should be treated as a subprocess (for instance, G_{21}, G_{12}).

To use the rule of relative gain:

The pair $i - j$ is correct if R_{ij} is closest to 1.

The rule can be appreciated if we examine the extreme case when either K_{12} or K_{21} equals zero.

Pairing Rule 3

It is also important to consider process dynamic behavior in determining the pairing.

1. A faster process should be paired as the main process such as G_{11} and G_{22}.
2. A slower process and processes with time delays should be treated as subprocesses (for instance, G_{21}, G_{12}).

Notes

1. If pairing rules 2 and 3 should result in a conflict, a trade-off is the only option.
2. Proper scaling may play an important role affecting the degree of difficulty of control, since it is related to both static gain and relative gain.
3. Correct pairing is one of the key issues that decide the quality of a MIMO control system. It is not unusual to see major improvements by simply correcting the pairing.

Degree of Interaction

Table 4.1 lists the control system design strategy based on the degree of interaction of the MIMO process. MFA controllers are always recommended.

Techniques for Adaptive Control

Table 4.1 MIMO System Design Strategy

Interaction Measure	Control Strategy
Small to no interaction, R is close to 1 ($0.9 < R < 1.1$)	Tighten both loops with SISO MFA.
Moderate interaction ($0.7 < R < 0.9$, or $1.1 < R < 2$)	Tighten important loops with SISO MFA and detune less important loops; or use MIMO MFA for better overall control.
Severe interaction ($0.5 < R < 0.7$, or $R > 2$)	Use MIMO MFA to control the process. May need to detune less important loops.

Decoupling and Compensator Gain

The MIMO model-free adaptive controller includes decoupling compensators. They are used to eliminate or reduce the control loop interactions.

Compensator Gain K_{c2}, K_{c3}

Similar to the design of MFA controller gain K_{c1}, the compensator gains K_{c2} and K_{c3} can be used to fine-tune the decoupling effects. The rule of thumb in tuning these gains is similar to the tuning of MFA controller gains, but should be set in a more conservative fashion. For instance, a K_{c2} should be set to only 0.5 if the estimated static gain of its related subprocess is 1.

ANTI-DELAY MFA CONTROL SYSTEM

In process control applications, many processes have large time delays due to the delay in the transformation of heat, materials, signals, etc. A good example is a moving strip process such as a steel rolling mill or a paper machine. No matter what control action is taken, its effect is not measurable without a period of time delay.

Since the time delay makes a time-invariant system time-varying, many linear time-invariant (LTI) system analysis tools such as root loci methods and LTI state-space equations cannot be used to study the processes with time delays.

Most importantly, a time delay causes the output to not respond to the control signal promptly. It is equivalent to disabling the feedback for a period of time. Feedback information is essential to automatic control. The measure of how significantly a time delay affects the process behavior is related to the $\tau{:}T$ ratio.

If a PID is used to control a process with significant time delays, the controller output will keep growing during the delay time and cause a large overshoot in system responses or even make the system unstable. Typically, a PID has to be detuned significantly in order to stay in automatic but will sacrifice control performance.

Generally speaking, a PID controller usually works for the process if its τ:T ratio is less than 1, unless it is detuned. When a controller is detuned, it loses the sharpness of its control capability so that the process cannot be tightly controlled.

A regular model-free adaptive controller works for the process if its τ:T ratio is less than 2. For the process with significant time delay (τ:T ratio greater than 2), special treatment is required.

A Smith Predictor is a useful control scheme to deal with processes with large time delays. However, a precise process model is usually required to construct a Smith Predictor. Otherwise, its performance may not be satisfactory.

SISO Anti-Delay MFA Controller

Figure 4.11 shows a block diagram for a single-input-single-output anti-delay model-free adaptive control system with an anti-delay MFA controller and a process with large time delays.

A special delay predictor is designed to produce a dynamic signal $y_c(t)$ to replace the process variable $y(t)$ as the feedback signal. Then, the input to the controller is calculated as

$$e(t) = r(t) - y_c(t) \qquad (4.54)$$

The idea here is to produce an $e(t)$ signal for the controller and let it "feel" its control effect without much delay so that it will keep producing proper control signals.

Since the MFA controller in the system has adaptive capability, the delay predictor can be designed in a simple form without knowing the quantitative information of the process.

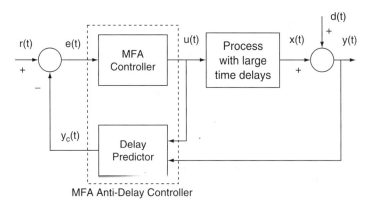

Figure 4.11   Anti-delay model-free adaptive control system.

172 Techniques for Adaptive Control

Compared to the traditional Smith Predictor, the anti-delay model-free adaptive controller does not need a precise process model. It only needs an estimated delay time as the basic information for its delay predictor. If the delay time used in the MFA delay predictor has a mismatch with the actual process delay time, the controller is robust enough to deal with the difference. Typically, it can deal with the situation where the delay time is 2 times larger or smaller than the actual delay time with satisfactory control performance. In addition, there is no real limitation on how large the $\tau:T$ ratio is allowed as long as a relatively close estimate of its delay time is provided.

Anti-delay MFA has achieved great success in real applications where processes with large and varying time delays such as electrical galvanizing lines are being effectively controlled. Anti-delay MFA's adaptive capability and special ability to deal with process time delays make it a valuable member of the MFA controller family.

SISO Anti-Delay MFA Configuration

To show how a practical anti-delay MFA controller is configured, we use an anti-delay controller configuration screen from CyboCon MFA control software as shown in Figure 4.12. Some key variables in this menu are described in the following:

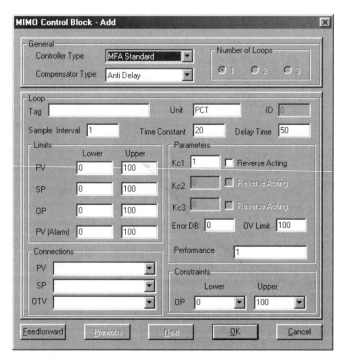

Figure 4.12   Anti-delay MFA controller configuration screen.

- *Delay Time.* The estimated delay time of the process in seconds. The time constant and controller gain can be configured like a SISO MFA controller.
- *Performance Index.* Based on the estimated process with K, T_c, and τ, let anti-delay MFA controller $K_c = 1/K$, time constant $= T_c$, and time delay $= \tau$. When the performance index $(I_p) = 1$, it is neutral. Increase I_p to tighten the control loop and decrease I_p to detune the controller.

MIMO Anti-Delay MFA Control System

Figure 4.13 illustrates a 2×2 multivariable anti-delay model-free adaptive control system. The anti-delay MFA controller set includes two MFA controllers C_{11} and C_{22}, two compensators C_{21} and C_{12}, and two predictors D_{11} and D_{22}. The process has large time delays in the main loops. Without losing generality, higher-order multivariable anti-delay MFA control systems can be designed accordingly. Because of the length of this chapter, a detailed description of this controller is not included.

MFA CASCADE CONTROL SYSTEM

When a process has two or more major potential disturbances and the process can be divided into two loops (one fast and one slow), cascade control can be used to take

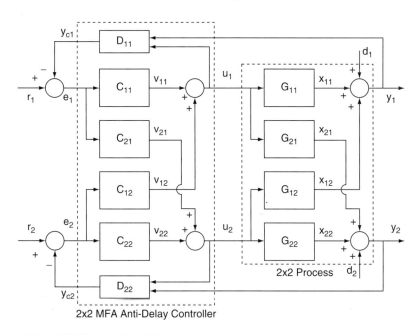

Figure 4.13 *MIMO anti-delay MFA control system.*

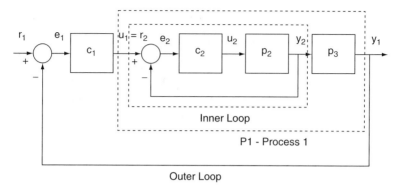

Figure 4.14 Model-free adaptive cascade control system.

corrective actions on the fast loop more promptly for better overall control performance. As illustrated in Figure 4.14, a cascade system contains two controllers, the primary controller C_1 and the secondary controller C_2. The inner loop consists of C_2 and P_2, and the outer loop consists of C_1 and P_1, where P_1 consists of C_2, P_2, and P_3. The output of C_1 drives the setpoint of C_2.

Although cascade control is one of the most useful control schemes in process control, it is often found that in real cascade control applications the outer loop is not closed.

Because of the interacting nature of the loops in the cascade control system, the requirement for proper controller tuning becomes much more important. If PI or PID controllers are used, four to six PID parameters have to be tuned. Good combinations of so many parameters are not easy to find. If the process dynamics change frequently, the controllers need to be retuned all the time. Otherwise the interacting nature of the inner and outer loop can cause serious system stability problems.

MFA in Cascade Control System

Using MFA controllers to form a cascade control system has an obvious advantage. Since the MFA controller used as C_2 can compensate for process dynamic changes, the closed-loop dynamics of the inner loop will change very little even though the process dynamics of P_2 may change a lot. This means the interconnection of the outer loop and the inner loop becomes much weaker. A more stable inner loop contributes to a more stable outer loop, and vice versa.

In addition, since each single-variable MFA controller has only one tuning parameter, the controller gain K_c, and it usually does not need to be tuned, the MFA based cascade control system becomes much easier to start up and maintain.

Multilayer Cascade Control System

Supervisory or setpoint control typically refers to a process optimization technique widely used in the process control industry. Basically, one calculates an optimal setpoint trajectory for the key process variable so that, if the process variable can track its setpoint trajectory, the process is optimized.

It is interesting to see that the cascade control system structure fits supervisory control quite well. The difference is that there is no need to derive a complex optimization algorithm to calculate the optimal setpoint trajectory; one can simply add a loop on the process variable that needs a "smart" setpoint trajectory. The question is, can this process variable be measured online?

For example, an industrial dryer typically has a cascade control system to control the dryer temperature using a temperature controller cascaded with a fuel flow controller. It is desirable to move the temperature setpoint around depending on how wet the raw material is and how large the load is. It is easy to imagine that calculating a setpoint trajectory for the temperature is not an easy task.

It is important to realize that the real control objective is almost always the quality variable. In this case, it could be the moisture of the product coming out of the dryer. Therefore, if we can measure the moisture of the product, we can simply add a control loop on top of the temperate and flow loop to make a three-layer cascade control system. The beauty of this design is that there is no need to calculate the complex optimal setpoint trajectory. The moisture controller's output is used as the setpoint trajectory of the temperature controller. The proper temperature control with an optimal setpoint trajectory will force the measured moisture to track its setpoint, which is determined based on the user's requirement or product specifications.

Automatic Control of Quality Variables

To conclude, a cascade control system has a unique importance to today's manufacturers where Six Sigma or zero defects quality objective is a necessity. Implementing a Six Sigma capable system can be very challenging. It is the author's belief that it is necessary to automatically control the quality variables such as density, moisture, product temperature, and dimension. Since there is usually a large delay time between the time when control action is taken and the time when the quality process variable is being measured, it is difficult to use a conventional controller such as a PID to control such quality variables. In this case, the anti-delay MFA controller is a good choice to be used as a quality variable controller with a multilayer cascade control system.

Cascade MFA Configuration

Configuration of a cascade control system using MFA controllers is relatively simple. According to the CyboCon software screen in Figure 4.15, it is seen that the Setpoint Connection field is filled with MFA-CAS1 (C1). Thus the setpoint of MFA-CAS2 (C2) is provided by the output of MFA-CAS1 when MFA-CAS2 is set in the remote mode.

There is no limitation on how many layers the cascade control system can include, as long as the process variables involved do not behave similarly. However, to simplify the system, it is recommended that you implement a cascade control system of only two or three layers.

The setup and tuning of the controllers in a cascade control system are carried out from the inside out. That is, the inner loop must be launched and working first; then the outer loop can be closed. It can be seen that the inner loop is part of the process for the outer loop.

The inner controller usually deals with a flow or pressure loop. If the process is relatively linear, it is not difficult to control since it is a relatively fast single-loop

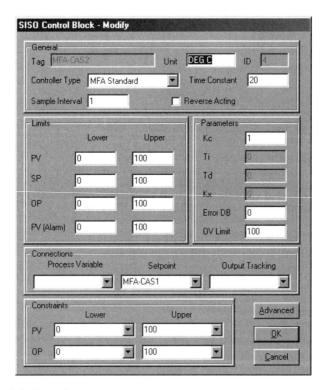

Figure 4.15 CyboCon software screen.

FEEDFORWARD MFA CONTROL

Basic Concept of Feedforward Control

Feedforward control, as the name suggests, is a control scheme to take advantage of forward signals. If (i) a process has a significant potential disturbance, and (ii) the disturbance can be measured; we can use a feedforward controller to reduce the effect of the disturbance on the loop before the feedback loop takes corrective action. If a feedforward controller is used properly together with a feedback controller, it can improve the control system performance quite economically.

Figure 4.16 illustrates a feedforward–feedback control system. The control signal $u(t)$ is combined with the feedback controller output $u_1(t)$ and the feedforward controllers output $u_2(t)$.

The feedforward controller is designed based on the so-called Invariant Principle. That is, with the measured disturbance signal, the feedforward controller is able to affect the loop response to the disturbance only. It does not affect the loop response to the setpoint change.

The control objective for the feedforward controller is to compensate for the measured disturbance. That is, it is desirable to have

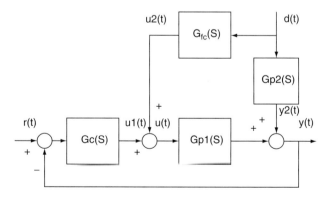

Figure 4.16 Feedforward and feedback control system.

178   Techniques for Adaptive Control

$$G_f(S) = \frac{Y(S)}{D(S)} = 0 \qquad (4.55)$$

where $G_f(S)$ is the Laplace transfer function of the feedforward loop; $Y(S)$ and $D(S)$ are the Laplace transform of process variable $y(t)$ and measured disturbance $d(t)$, respectively. Then, the FF controller can be designed as

$$G_{fc}(S) = -\frac{G_{p2}(S)}{G_{p1}(S)} \qquad (4.56)$$

where $G_{fc}(S)$ is the Laplace transfer function of the feedforward controller.

Feedforward compensation can be as simple as a ratio between two signals. It could also involve complicated energy or material balance calculations.

Feedforward MFA Controller

For ease of use, it is desirable to design a more general-purpose feedforward controller so that it can be easily configured and launched.

The feedforward MFA controller is designed to be such a general-purpose feedforward controller that works with the feedback MFA controllers. The idea is not to attempt a perfect cancellation to the disturbances. This is because Invariant Principle–based perfect disturbance cancellation is very difficult to implement in industrial applications. When the feedback controller is MFA, its adaptive capability just makes many conventional control methods easier to implement and more effective.

There is no need to find the process models for $G_{p2}(S)$ and $G_{p1}(S)$ to design the feedforward controller $G_{fc}(S)$. Configuration, commissioning, and maintenance of the FF MFA controller is much simpler than for a regular FF controller.

Feedforward MFA Configuration

To show how a feedforward MFA controller is configured, we use a configuration screen from CyboCon MFA control software as shown in Figure 4.17. Some key variables in this menu are described in the following:

FF Gain K_{fc}

It is important to enter the correct sign for the gain, which can be found based on the following formula:

$$K_{fc} = -\frac{K_{p2}}{K_{p1}} \qquad (4.57)$$

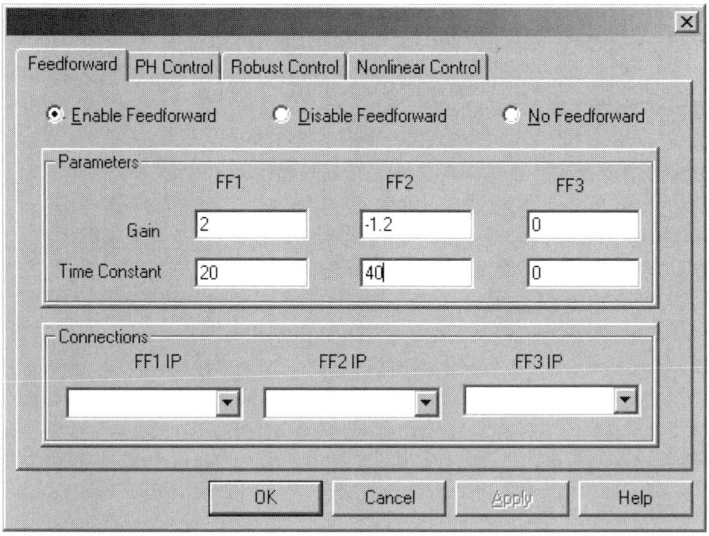

Figure 4.17 Configuration screen.

where K_{fc} is the gain for the Feedforward controller, and K_{p1} and K_{p2} are the estimated static gain for processes G_{p1} and G_{p2}, respectively.

In order to ensure that the feedforward action rejects the disturbance, the rules of selecting the sign can be summarized as follows:

- If G_{p1} and G_{p2} have the same sign, the FF's gain should be negative.
- If G_{p1} and G_{p2} have different signs, the FF's gain should be positive.

FF Time Constant

The Time Constant field can be filled with an estimate of the time constant of G_{p2}. This is related to how fast the disturbance will affect the process PV.

Feedforward MFA for Process with Large Time Delays

A process with large time delays can be complicated when applying a feedforward controller. As shown in Figure 4.18, the potential time delays can be part of G_{p1} and G_{p2} or in a separate process with pure time delay DT. Because of the length of this chapter, we will not discuss this in detail. Interested users can refer to Cheng (2000 or 2001a) for more information.

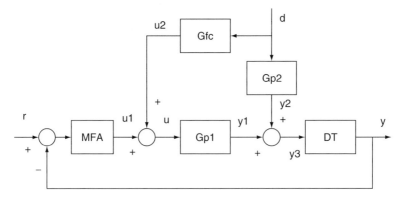

Figure 4.18   Feedforward for process with large delay time.

NONLINEAR MFA CONTROL

Nonlinear MFA Controller

Nonlinear control is one of the most challenging topics in modern control theory. Whereas linear control system theory has been well developed, nonlinear control problems present many headaches.

The main reason that a nonlinear process is difficult to control is because there could be so many variations in process nonlinear behavior. Therefore, it is difficult to develop a single controller to deal with the various nonlinear processes.

Traditionally, a nonlinear process has to be linearized first before an automatic controller can be effectively applied. This is typically achieved by adding a reverse nonlinear function to compensate for the nonlinear behavior so that the overall process input/output relationship becomes somewhat linear. It is usually a tedious job to match the nonlinear curve; and process uncertainties can easily ruin the effort.

A nonlinear MFA controller has been developed to deal with a wide range of nonlinear processes. It provides a more uniform solution to nonlinear control problems. The nonlinear MFA controller is well suited for (i) nonlinear processes and (ii) the processes with nonlinear actuators. A high-pressure loop is a typical nonlinear process that can cause the actuator to lose its authority in different operating conditions. Inevitable wear and tear on a valve typically makes a linear valve nonlinear.

Nonlinear MFA Configuration

The screen in Figure 4.19 shows how this nonlinear MFA controller is configured.

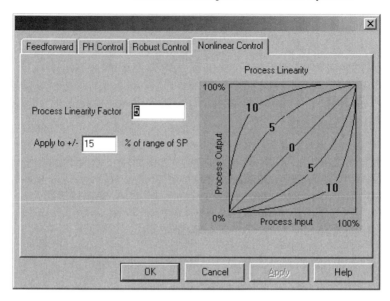

Figure 4.19   Nonlinear MFA controller configuration screen.

Process Linearity Factor

The graph in Figure 4.19 shows how severe the nonlinear behavior is between the process input and process output. The process linearity factor is a number between 0 and 10. A 10 represents an extremely nonlinear process while a 0 represents a linear process. Notice that the graph shows a nonlinear curve marked with 10 on both upper and lower positions. This means a nonlinear MFA controller does not care what the nonlinear characteristics are for this process. For instance, the valve can be either "fast open" or "fast close" as represented by these two convex and concave curves.

When using nonlinear MFA, you do not need to worry about how the nonlinear curve is laid out. It is your option to tell the controller whether the process is extremely nonlinear (give a 9 or 10), quite nonlinear (give a 5 or 6), or somewhat nonlinear (give 1 or 2). The nonlinear MFA controller will be smart enough to handle the rest.

Simulations and real applications show that the nonlinear MFA controller can easily deal with a nonlinear process even if its gain changes hundreds of times. For a nonlinear MFA, there is no linearization calculation or process model. The MFA controller gain K_c is simply set at its nominal point and not retuned.

To conclude, the nonlinear MFA controller provides effective and tight control of flow, pressure, and other key process variables with consistent performance.

Less loop oscillation and smoother operation result in higher product quality, improved production efficiency, and less energy and material consumption. Since it can effectively control the flow and pressure in its entire operating range without the need to retune its parameters, this controller is especially useful for the multilayer cascade control system as the inner controller. It improves the overall control system stability and enables automatic control of the quality variable that is controlled by the outer loop controller.

MFA pH Controller

Most process plants generate a wastewater effluent that must be neutralized prior to discharge or reuse. Consequently, pH control is needed in just about every process plant, yet a large percentage of pH loops perform poorly. Results are inferior product quality, environmental pollution, and material waste.

With ever-increasing pressure to improve plant efficiency and tighter regulations in environmental protection, effective pH control is very desirable. However, implementing a pH system is like putting a puzzle together. It will only work when all the components are in place.

The pH puzzle includes effective pH probes, actuators, and controllers. Whereas various pH probes and actuators for pH control are available, commercial adaptive pH controllers are still scarce.

The MFA pH value controller is able to control a wide range of pH loops because its powerful adaptive capability allows it to compensate for the large nonlinear gain changes. It controls full pH range with high precision and enables automatic control of acid or alkaline concentration, which are critical quality variables for chemicals.

MFA pH Configuration

A SISO MFA controller can be configured with special capability for controlling strong-acid-strong-base pH loops. As shown in Figure 4.20, you can enter the break points A and B to define the estimated shape of the titration curve of the pH process. Then you can enter the MFA controller gain Kc for the flat portion and steep slope, respectively. Because of the adaptive capability of the MFA controller, the titration curve does not have to be accurate and, in fact, its shape can vary in real applications.

MFA pH control has helped many users to effectively control their tough pH loops. Quick return-on-investment was reported with savings on chemical reagents, no violation of discharge code, and smoother production operation.

Model-Free Adaptive Control with CyboCon

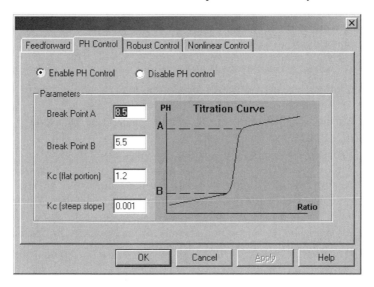

Figure 4.20 MFA pH controller configuration screen.

ROBUST MFA CONTROL

In complex control applications, we may face the following challenges that even a regular MFA controller cannot effectively handle:

1. There is a big change in the system dynamics so that a regular MFA controller is unable to provide prompt control action to meet the control performance criteria;
2. The dominant disturbance to the system cannot be economically measured so that feedforward compensation cannot be easily implemented;
3. A controller purposely detuned to minimize the variations in its manipulated variable may lose control when there is a large disturbance or significant dynamic behavior change; or
4. The system dynamic behavior or load change does not provide triggering information to allow the control system to switch operating modes.

A Chemical Reactor Example

To describe the application in more detail, a chemical reactor control problem is studied in the following. Chemical batch reactors are critical operating units in the chemical processing industry. Controlling the batch reaction temperature is always a challenge because of the complex nature of the process, large potential disturbances, interactions between key variables, and multiple operating conditions. A large percentage of batch reactors running today cannot keep the reactor temperature in automatic control throughout the entire operating period, thus resulting in lower efficiency, wasted manpower and materials, and inconsistent product quality.

An exothermal batch reactor process typically has four operating stages:

1. *Startup stage*. Ramps up the reactor temperature by use of steam to a predefined reaction temperature.
2. *Reaction and holding stage*. Holds the temperature by use of cooling water while chemical reaction is taking place and heat is being generated.
3. *No-reaction and holding stage*. Holds the temperature by use of steam after main chemical reaction is complete and heat is not being generated.
4. *Ending stage*. Ramps down the reactor temperature for discharging the products.

During the transition period from stage 2 to stage 3, the reactor can change its nature rapidly from a heat-generation process to a heat-consumption process. This change happens without any triggering signal because the chemical reaction can end at any time depending on the types of chemicals, their concentration, catalyst, and reaction temperature. Within a very short period of time, the reactor temperature can drop significantly. The control system must react quickly to cut off the cooling water and send in a proper amount of steam to drive the reactor temperature back to normal. A regular feedback controller is not able to automatically control a batch reactor during this transition if it is tuned to control the process in stages 1 and 2. In practice, batch reactors are usually switched to manual control and rely on well-trained operators during critical transitions. It is a tedious and nerve-wracking job that can result in low product quality and yield.

Robust MFA Controller

A robust MFA controller is developed to control the problematic processes described. Robust control is usually referred to as a controller design method that focuses on the reliability (robustness) of the control algorithm. Robustness is defined as the minimum requirement a control system has to satisfy to be useful in a practical environment. Once the controller is designed, its parameters do not change and control performance is guaranteed.

Robust MFA is not a control system design method. We use the term "robust" here because this novel controller is able to dramatically improve the control system robustness. Without the need to redesign a controller, use feedforward compensation, or retune the controller parameters, the robust MFA controller is able to keep the system in automatic control through normal and extreme operating conditions when there are significant disturbances or system dynamic changes.

Model-Free Adaptive Control with CyboCon

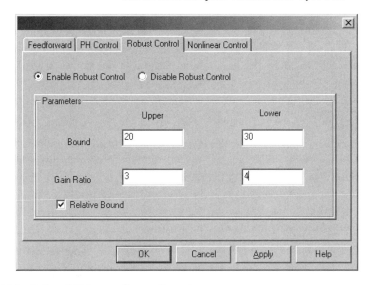

Figure 4.21   Robust MFA controller configuration screen.

Robust MFA Configuration

As shown in Figure 4.21, the robust MFA controller can be easily configured with these parameters.

Upper and Lower Bound

These are the bounds for the process variable (PV) being controlled. They provide intelligent upper and lower boundaries for the process variable. These bounds are typically the marginal values that the process variable should not go beyond.

PV is unlike OP where a hard limit or constraint can be set. PV is a process variable that can only be controlled by manipulating the OP. Therefore, the upper and lower bounds for PV are very different from the OP constraints.

For instance, if you have a level loop with a setpoint of 50 and you want to have a $+20$ and -30 bound around the setpoint, you should enter upper bound $= 20$, lower bound $= 30$.

Relative Bound

If this box is checked, the upper and lower bounds you enter are related to the setpoint as described above. If this box is not checked, the upper and lower bounds you enter are the actual bounds.

Gain Ratio

The gain ratio is the coefficient to increase or decrease the MFA control action. Since MFA controller gain K_c is the only tuning parameter we use, we use the so-called gain ratio to relate this parameter to the actual MFA gain K_c. Typically, you want to enter Gain Ratio = 3, which implies that the MFA gain working in the abnormal situation is 3 times higher than the regular MFA gain setting.

It is important to understand that this is not a gain scheduling approach, although it appears to be this way. Gain scheduling will not be able to resolve the complex problems described.

MFA CONTROL METHODOLOGY AND APPLICATIONS

The concept, architecture, algorithm, system structure, and configuration of various model-free adaptive control systems were described in the previous section. To have a deeper understanding of this unique control method, it is necessary to (i) study the philosophy and methodology used in developing these MFA control systems, and (ii) review the MFA control applications that have proven the concept.

MFA CONTROL METHODOLOGY

"All roads lead to Rome."

A problem usually has multiple possible solutions. A process can usually be controlled using different controllers based on different control methods. Almost every control method has its merits and weakness. What's important is to develop the right type of controller to fit the right type of application at a minimum cost.

In natural science, the combination of physics, mathematics, and philosophy plays an integral part in developing a theory that is practically useful. Physics is the foundation for the study of the physical process or environment; mathematics provides the tools to precisely describe the physical process or phenomenon; and, equally important, philosophy provides directions.

The development of model-free adaptive control technology started from a simple desire to develop a new controller that could easily and effectively solve various industrial control problems. The actual development process has evolved from a prolonged interest in the study of combined intelligence methodology. Since model-free adaptive control does not follow the traditional path of model-based adaptive control, the philosophy behind combined intelligence has led the way up this long and rocky road.

Combined Intelligence Methodology

The combined intelligence methodology developed by the author of this chapter consists of the following problem-solving philosophy:

1. Always seek a simple solution for a complex problem
2. Use all information available
3. Do not depend on the information's accuracy
4. Apply a technique that fits the application

These four key points are described in the following.

Seek a Simple Solution

A simple solution is almost always the best solution. A complex solution might achieve a little better result, but the cost can be very high. Most users want to have a tool or system that is easy to use, launch, and maintain with the best price–performance ratio. A simple solution usually fits this need well.

Use All Information Available

It is a cliché, but we are living in the information era. Information has value. A small piece of information can make all the difference. When solving a problem, do not waste the valuable information available. For instance, a process delay time can easily be seen from the process trend chart. A regular PID controller ignores this important piece of information.

Do Not Depend on the Information's Accuracy

Practically, information received may not be accurate. What's worse is that we often do not even know whether the information is accurate or not. If we knew, we would simply have the option to use or not use the information. For this reason, the solution we provide has to be agile or adaptive enough that it can deal with inaccuracy of the information and with uncertainties.

Apply a Technique That Fits the Application

Arguments often arise between people who believe in very different problem-solving methods. For instance, model-based and rule-based methods are two very distinctive approaches in control theory. Since almost all methods have their merits and shortcomings, why argue? Let's use the technique that fits the application.

The MFA Control Approach

To see how the MFA control method is developed based on this problem-solving philosophy, we will relate MFA to each one of the points in the following discussion.

Seek a Simple Solution

PID control is simple since it is a general-purpose controller and its algorithm is easy to understand. However, PID is almost too simple to control complex systems. In this regard, PID cannot be considered an effective solution to the more difficult control problems. On the other hand, model-based advanced control methods have proven themselves too complex to launch and maintain since they depend on either a first principle or an identification-based process model. A dream controller has to be powerful enough to control various complex processes yet simple enough to use, launch, and maintain. MFA is a solution that fits these requirements.

Use All Information Available

Model-free adaptive control, as its name suggests, is a control method that does not depend on either first principle or identification-based process models. However, we do try to use all the process information available. For this reason, it can be considered an information-based controller.

For instance, the process time constant defines how fast a dynamical system responds to its input. A slow process might have a 10-hour time constant and a fast process might have a 10-millisecond time constant. It would be unwise not to use this information for the controller. In addition, it is relatively easy to estimate the time constant by reading a trend chart. Other important yet easily obtained information about a process includes its acting type (either direct or reverse), static gain, and delay time, if any. As described earlier, an MFA controller is designed to use the process parameters that can be easily estimated.

Do Not Depend on the Information's Accuracy

A process can be classified as a white, gray, or black box. If its input/output relationship is clear, the process is a white box. We can easily use existing well-established control methods and tools to design a controller for this process.

When we are not sure if the process input/output relationship is accurate, or if the process has potential disturbances, dynamic changes, and uncertainties, the process is a gray box. In this case, MFA's adaptive capability is able to handle such changes and uncertainties. PID or model-based control methods will have a much tougher time or higher cost addressing these uncertainties.

Apply a Technique That Fits the Application

MFA is neither model-based nor rule-based. We might say that it is an information-based control method. If the argument is made that the process information used is equivalent to a process model, that's perfectly acceptable. The key to our approach is that we focus on delivering a simple, adaptive, and effective solution.

To extend this idea, a series of MFA controllers have been developed to address different difficult control problems. Users can simply pick the appropriate MFA, configure its parameters, and launch the controller.

THE INSIDE OF MFA

Since its introduction, the concept of model-free adaptive (MFA) control has puzzled many people. The very definition of MFA control makes strong claims and it is difficult to comprehend how it works. Although detailed architecture and algorithm of MFA controllers were presented earlier, the following discussion summarizes what MFA control is and how it works.

A comprehensive article about model-free adaptive control was published in *Control Engineering Europe* magazine's February/March 2001 issue (VanDoren, 2001).

What MFA Is Not

MFA is not

- A self-tuning PID controller
- A model-based adaptive controller
- A model-based neural network controller
- A fuzzy controller

What MFA Is

MFA is a novel, patented, and "model-free" adaptive controller. MFA consists of a nonlinear dynamic block that performs the tasks of a feedback controller. A dynamic block is just a dynamic system with inputs and outputs. The control objective is to produce an output to minimize the error between the setpoint and the process variable (PV) being controlled.

Within the dynamic block there is a group of weighting factors that can be changed as needed to vary the nonlinear functions of the block. Then the MFA controller can adapt as the process dynamics change. In addition, the MFA controller "remembers" a portion of the process data enabling it to make quick and precise changes to its output.

Why "No-Model" Is Possible

Automatic control methods are all based on the concept of feedback. The essence of feedback theory consists of three components: measurement, comparison, and correction. Measuring the quantity of the variable to be controlled, comparing it with the desired value, and using the error to provide the control action is the basic procedure of automatic control.

The key to the theory and application of automatic control is the ability to make the correction based on proper measurement and comparison. An automatic controller does not require a process model to produce an output. Therefore, an adaptive controller does not have to rely on a process model to produce an output.

Why Use a Model?

A feedback automatic controller relies on the feedback signal to produce its output and does not require a process model to produce an output. That means there should be a way to compute control actions directly from the process input and output signal without first creating a model to mimic the process behavior. The basic process information such as time constant is available and the dynamical information that involves the changes and uncertainties is included in the process input and output data. That means we should be able to "crunch" the data correctly to let the controller generate a proper output.

In other words, computing control action does not require a process model. The real trick is to compute the control action and vary the weighting factors based on the input/output data. The key to MFA is a set of algorithms that can do this job.

MFA System Stability

In model-free adaptive control theory, a sufficient condition for MFA control system stability is derived and can be described as follows: If the process is passive (such as open-loop stable), the closed-loop MFA system stability is guaranteed and the process can be linear, nonlinear, time-invariant, time-varying, single-variable, or multivariable.

MFA Requirements

1. *The process is controllable.* If it is not controllable, you need to change the process or improve the variable pairing.
2. *The process is open-loop stable.* If it is not open-loop stable, it must first be stabilized. However, for certain simple open-loop unstable processes such as a nonself-regulating level loop, no special treatment is required.
3. *The process is either direct or reverse acting (the process does not change its sign).* If a process changes its sign within its operating range, it is still possible to design a special controller for this "ill" process.

Key Issues of MFA

- *MFA does not include an identification mechanism.* No dynamic modeling mechanism is used in MFA. In other words, there is no identification engine inside MFA. This approach eliminates the headaches that a model-based control method causes, such as offline model training or the possibility of obtaining a bad model through identification. For instance, when a sensor fails, an identification mechanism may learn a bad process model, which can jeopardize the control system.
- *MFA adapts as is needed.* Thus, MFA does not need continuous excitation of the process signals. In other words, if the process dynamics do not change, the controller will not attempt to vary the weighting factors.
- *MFA is a robust controller.* When MFA is installed, the adaptation will not cause it to develop a strange behavior because it does not rely on a model. In addition, since it is an adaptive controller, its robust range is wider than that of a regular PID, which does not adapt at all.
- *MFA is not a single solution.* Based on the core of MFA, different types of MFA controllers have been developed to deal with special problems. For instance, anti-delay MFA is very effective in controlling processes with large time delays; and nonlinear MFA is well suited to control nonlinear processes such as pressure loops where valves may lose authority. More special MFA controllers are being developed as needed.

CASE STUDIES

So far, we have discussed the ideas of MFA control in theory. Let's see how it works in practical applications. Five application stories were selected that discuss different types of MFA controllers, including SISO, MIMO, anti-delay, nonlinear pH, and robust MFA.

SISO MFA on Tomato Hot Breaks

The tomato hot-breaks temperature control application is at Del Monte Foods' Woodland, CA, plant, and an article entitled "Adaptive and Predictive Controls Boost Product Quality" describing this application was published in *Food Engineering Magazine*'s December 1999 issue (Morris, 1999).

From July through early October, the plant operates 24 hours per day as a continuous caravan of gondola trucks unloads tomatoes into flumes feeding the hot break lines. Tomatoes are chopped and fed into heating vessels to become tomato slurry. Production throughput is critical to cost efficiency during the short processing season. The major challenge is to maintain good temperature control of these rotary-coiled heating vessels. Tomato inflow can be wild between truckloads, causing large temperature variations. Too high a temperature causes burning and clogging, and too low a temperature affects the production efficiency of the downstream

Techniques for Adaptive Control

Figure 4.22   MFA control station and tomato lines.

evaporators. In addition, overheating the tomato slurry wastes energy and causes quality problems. Figure 4.22 illustrates the MFA control station and the tomato lines used at Del Monte's Woodland plant.

If the tomato flow could be measured, feedforward control would be applied to solve the control problem. Measuring the solidity of tomatoes is difficult and costly. Most tomato processing plants are not equipped to measure solids. PID controllers were previously used to control these nine heating vessels where temperature swings typically varied as much as 15 to 20°F.

The plant installed CyboCon model-free adaptive (MFA) control software with nine SISO MFA controllers to control the temperature of the heating vessels. CyboCon was installed in just a few hours. The PID loops were retained offering the operator a choice of control, "but since installation the operators have used CyboCon 100% of the time," according to Del Monte's plant operations manager. Product temperature now typically varies within less than ±2°F.

Figure 4.23 shows that the MFA controllers in CyboCon software quickly and tightly control the temperature by manipulating steam to compensate for wild tomato inflow without using feedforward control.

Since its installation, no maintenance has been required to retune these controllers. The plant installed more MFA controllers the following season to control boilers and steam injection systems for sterilization, which all achieved the desired control objectives and economic benefits.

MIMO MFA on Multizone Temperature Control

A multizone temperature MFA control for industrial coking furnaces is installed at the Guangzhou Petrochemical Complex of SINOPEC in China, and an article

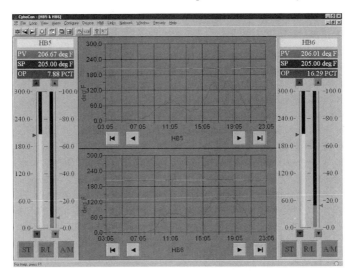

Figure 4.23 MFA control screen.

entitled, "Model-Free Coking Furnace Adaptive Control" presenting the application was published in *Hydrocarbon Processing* magazine's December 1999 issue (Cheng et al., 1999).

Control of temperature loops in multiple zones can be problematic, especially when a narrow specification is required. Successful installation of a model-free adaptive (MFA) control system in the delayed coking process at the Guang-Zhou Petrochemical Complex (GPC) shows how this problem can be resolved.

As shown in Figure 4.24, a coker consists of two coking furnaces, each with two combustion chambers. High temperatures create carbon that clogs pipes, and a below-spec temperature causes an insufficient reaction so that the yield drops.

Control difficulties result from (i) large time delays, (ii) serious coupling between these two temperature loops because the separation wall between the two combustion chambers is quite low, and (iii) multiple disturbances in gas pressure, oil flow rate, oil inflow temperature, and oil composition. The oil outlet temperature is sensitive to the gas flow rate change, and the temperature specification is tight ($\pm 1^\circ$C).

A 2×2 anti-delay MFA controller was installed to manipulate the fuel flow valves on each furnace to control the outlet temperature. The MIMO anti-delay MFA solved large time delay and coupling problems. MFA controllers compensated for disturbances and uncertainties. Constraints on controller outputs prevented temperatures from running too high or too low.

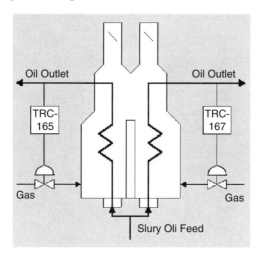

Figure 4.24   Control diagram of coking furnace.

According to the chief engineer at GPC, MFA controllers started automatic control with no bumps to the system. Commissioning took 3 days, with the following results: (i) both furnaces are automatically controlled under all conditions; (ii) outlet oil temperature is controlled to within $\pm 1°C$ with energy savings and consistent product quality; (iii) operators have been relieved of tedious, ineffective manual control responsibilities; and (iv) higher efficiencies and yields have been achieved.

Anti-Delay MFA on Metal Galvanizing Process

This anti-delay MFA control application is at the rolling mill of Wuhan Iron & Steel Corporation in China.

In a rolling mill, a galvanizing process produces tinned plating. Plating thickness is key to product quality. Plating that is too thick or thin causes an economic loss. If the layer is too thick, precious metal is wasted; and if it is too thin, defective parts are produced.

The tinning processing line of Wuhan Iron & Steel (Group) Co. (WISCO) performs large-scale continuous, high-speed tinned plating. Usually, on most plating lines, thickness can only be controlled manually. Differences in operator skills led inevitably to inconsistent product quality.

As shown in Figure 4.25, to keep plating continuous, buffers that store strip up to 150 meters long are arranged on both sides of the galvanizing tank. The outlet buffer stores strip when cutting while the inlet buffer supplies strip when welding.

Plating thickness could only be controlled manually, and it was manipulated by the electrical current. The amount of current applied was calculated according to the

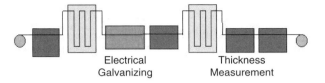

Figure 4.25 Galvanizing line.

thickness, width, and speed of strip. Concentration and temperature of the galvanizing liquid are other variables that can affect the thickness. Large disturbances in strip speed affected thickness even though the current was adjusted based on a process model.

Behind the outlet buffer, a sensor with a range of $0-15\,g/m^2$ measures plating thickness online. This provides an opportunity to control this quality variable automatically. However, the buffer between the galvanizing tank and the sensor creates a large and random time delay that changes from 30 to 150 sec and cannot be easily predicted. A variation in sizes of products (more than 20 types per day) causes frequent changes in process dynamic behavior and time delay. Other disturbances affect product quality and also make automatic control extremely difficult.

Two anti-delay MFA controllers were installed to control the plating thickness of the top and bottom layers. A feedforward MFA controller for each feedback MFA was also used to quickly overcome the speed disturbance. MFA control succeeded in reaching the objectives defined by WISCO:

1. *Control of plating thickness to within* $\pm 0.5\,g/m^2$. MFA controllers maintain thickness within $\pm 0.3\,g/m^2$ even if strip speed changes severely; and
2. *Ease of installation and operation.* Soon after connecting to the PLC, MFA controllers for both top and bottom layers were launched and controlled plating thickness immediately. Commissioning was completed in 2 days. Return-on-investment was achieved in a short period of time because of automatic control of the quality variables.

MFA pH Control on Water Neutralization Process

This MFA pH control application is at the Rohm & Haas Company in Houston, Texas. Rohm and Haas, a leading chemical company, is successfully using an MFA pH controller to control a problematic pH loop to neutralize an organic process stream with an estimated savings at $170,000 per year. A reliability improvement is also achieved due to reduced formation of solids, according to the control engineer at Rohm & Haas.

The stream to be neutralized is a two-phase stream with varying concentrations of acidic species. Caustic water is added as reagent to neutralize the stream. In this case, pH control is difficult because of the strong nonlinearity of the pH process and the large load changes in the stream inflow.

The original system was designed with one caustic valve manipulated by a single-loop PID controller. The control performance was not satisfactory. One tedious solution might involve reengineering the process and/or using the process data to create a mathematical model to linearize the process input and output relationship.

The recommended pH setpoint was 10.6. However, operators typically ran the process at 12 because the pH loop became unstable when it was close to the recommended setpoint. Excess caustic from the higher pH resulted in solids formation in the downstream separation equipment. The result was wasted chemical material and excessive clogging. On the other hand, setting the PID gain low to improve control stability would result in an extremely sluggish control response when a large upset pushed the pH too far away from the neutrality region.

As illustrated in Figure 4.26, a CyboCon CE model-free adaptive control instrument was installed to control this pH process. The CyboCon CE was mounted on the existing panel and wired directly to the process. Launching the MFA pH controller was simple with no complicated tuning, step testing, or data collection involved.

Improved pH control enabled Rohm & Haas to lower the pH setpoint from 12 to 11. Not only were cost benefits achieved, operators also liked the improved process

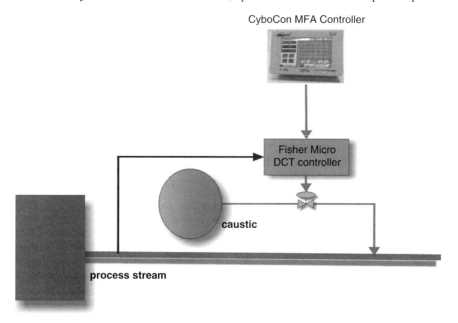

Figure 4.26 MFA pH control diagram.

upset handling capabilities. In addition, reduction of excess caustic and reduced solids formation meant an unquantified improvement in overall system reliability.

Other MFA pH control applications, including the ones at Chiron in California and Ultrafertil in Brazil, achieved similar results within a short return-on-investment period.

Robust MFA on Distillation Column

Robust MFA controllers were installed in two Air Liquide America locations, and an article entitled "Air Separation Advances with MFA Control" describing the details of these applications was published in *Control Magazine*'s May 2001 issue (Seiver and Marin, 2001).

Air Liquide America, a global provider of industrial, electronic, and health care gases, has standardized on model-free adaptive control for advanced regulatory control applications after successful MFA installation on two air separation units (ASUs).

Figure 4.27 illustrates the process diagram of an air separation unit. The main goal of operating an ASU is to maximize yields of gases including oxygen, nitrogen, and argon; and maintain the operation in as steady a state as possible. The specific goal of

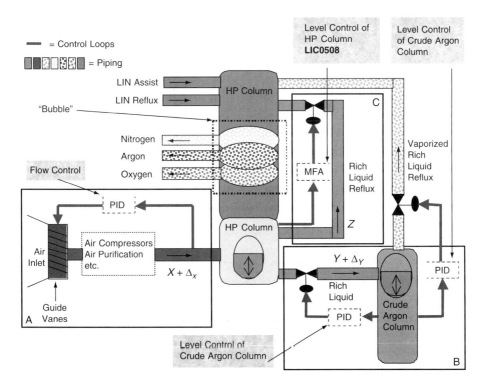

Figure 4.27   Diagram of an air separation unit.

the initial application was to control the rich liquid (RL) reflux level in the high pressure (HP) cryogenic column so that it would remain as constant as possible, even during plant ramping and upsets. The RL reflux flow to the low-pressure (LP) cryogenic column is used to manipulate the HP column RL reflux level.

It is difficult to properly tune a PID controller for good control performance under all conditions on an ASU because of the variable rates of the HP column inflow and outflow. Overly tight level control will result in large oscillations in the reflux flow, which causes a lower product yield. A PID level controller is usually detuned to allow the level to fluctuate to minimize variations of the reflux flow. This may result in safety problems during a plant upset since the detuned PID cannot deal with large disturbances. In addition, oscillations in level can cause the process to swing, which also results in a lower yield.

The robust MFA controller has been used to tightly control the level. A surprising result is that both level and reflux variations have been reduced at the same time. MFA immediately started to set production records as soon as it came on line. MFA control proved quite easy to install on ASUs. Air Liquide staff engineers performed the entire installation and commissioning at their McMinnville, Oregon, plant within a single day. Since its installation, virtually no maintenance or retuning has been required.

Figures 4.28 to 4.31 show the trend charts obtained from the actual system at Air Liquide that compare the performance of PID and MFA control.

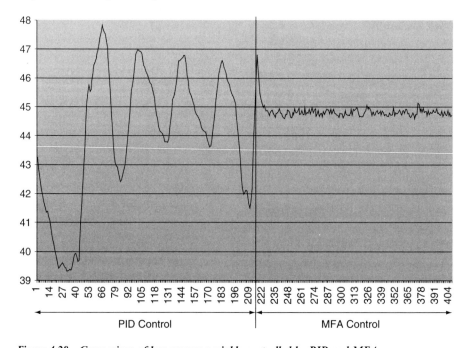

Figure 4.28 Comparison of key process variable controlled by PID and MFA.

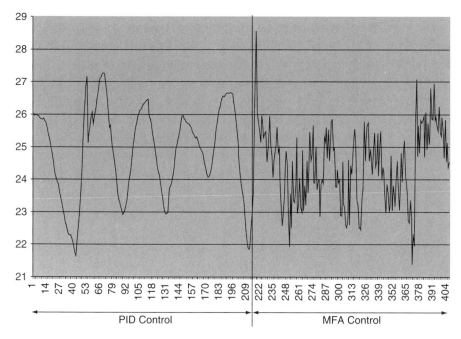

Figure 4.29   Reflux flow change from PID control to MFA control.

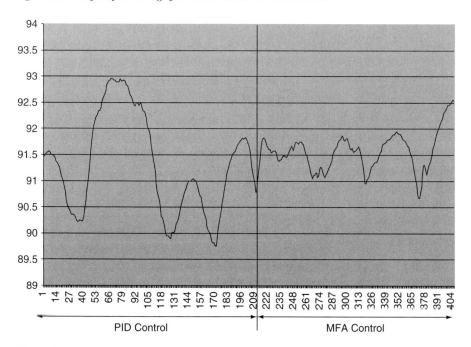

Figure 4.30   Principal column purity when switching from PID to MFA control.

200 Techniques for Adaptive Control

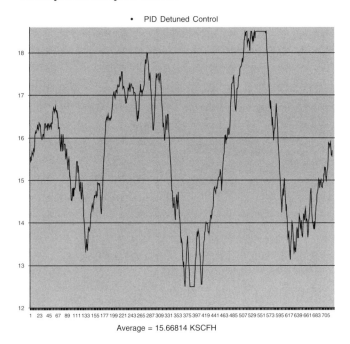

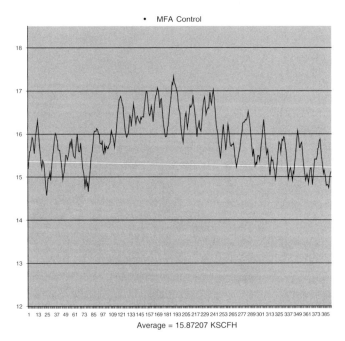

Figure 4.31 Improvement in gas flow yield.

According to Air Liquide's advanced control manager, by using model-free adaptive control, Air Liquide achieved benefits in the areas of product yield, quality control, and, most importantly, operational stability.

REFERENCES

Anderson, B. D. O., et al. (1986). *Stability of Adaptive Systems: Passivity and Averaging Analysis.* MIT Press, Cambridge, MA.

Cheng, George S. (1996). *Model-Free Adaptive Control Theory and Applications.* A dissertation for the doctor of engineering degree, Shanghai Jiao-Tong University, Shanghai, China.

Cheng, George S. (2000). *MFA in Control with CyboCon.* CyboSoft, General Cybernation Group Inc., Rancho Cordova, CA.

Cheng, George S. (2001a). *CyboCon and Model-Free Adaptive Control Training Manual.* CyboSoft, General Cybernation Group Inc., Rancho Cordova, CA.

Cheng, George S., Min He, and De-Lin Li (1999). "Model-Free Coking Furnace Adaptive Control," in *Hydrocarbon Processing.* Gulf Publishing Company, Houston TX, December 1999.

Desoer, C.A., and M. Vidyasagar (1975). *Feedback Systems: Input–Output Properties.* Academic Press, New York.

Hang, Chang C., Tong H. Lee, and Weng K. Ho (1993). *Adaptive Control.* Instrument Society of America, Research Triangle Park, NC.

Morris, Charles (1999). "Adaptive and Predictive Controls Boost Product Quality," *Food Engineering,* December.

Murrill, P. (1981). *Fundamentals of Process Control Theory.* Instrument Society of America, Research Triangle Park, NC.

Seiver, Dave, and Ovidiu Marin (2001). "Air Separation Advances with MFA Control," *Control,* May.

Shinskey, F. G. (1979). *Process-Control Systems,* McGraw-Hill Book Co., New York.

VanDoren, Vance (2001). "Model Free Adaptive Control—This New Technique for Adaptive Control Addresses a Variety of Technical Challenges," *Control Engineering Europe,* March.

SUGGESTED READING

Åström, K. J. (1984). *Computer Controlled Systems, Theory and Design.* Prentice-Hall, Inc., New York.

Åström, K. J., and Tore Hagglund (1988). *Automatic Tuning of PID Controllers.* Instrument Society of America, Research Triangle Park, NC.

Åström, K. J., and B. Wittenmark (1989). *Adaptive Control.* Addison-Wesley, New York.

Athans, Michael, and Peter Falb (1966). *Optimal Control,* McGraw-Hill Book Co., New York.

Bennett, S., and D. A. Linkens, editors (1984). *Real-Time Computer Control.* Peter Peregrinus Ltd., London.

Buzzard, W. S. (1994). *Flow Control.* Instrument Society of America, Research Triangle Park, NC.

Cheng, George S. (2001b). *CyboCon CE MFA Control Instrument User Manual.* CyboSoft, General Cybernation Group Inc., Rancho Cordova, CA.

Clarke, David, editor (1994). *Advances in Model-Based Predictive Control.* Oxford University Press, Oxford.
Corripio, A. B. (1998). *Design and Application of Process Control Systems.* Instrument Society of America, Research Triangle Park, NC.
Corripio, Armando (1990). *Tuning of Industrial Control Systems.* Instrument Society of America, Research Triangle Park, NC.
Deshpande, Pradeep, editor (1989). *Multivariable Process Control.* Instrument Society of America, Research Triangle Park, NC.
Fu, K. S. (1970). "Learning Control Systems—Review and Outlook," *IEEE Trans. Autom. Control* **16**, 210–221.
Goodwin, Graham, and Kwai Sang Sin (1984). *Adaptive Filtering, Prediction, and Control.* Prentice-Hall, Inc., New York.
Gunkler, A. A., and J. W. Bernard (1990). *Computer Control Strategies for the Fluid Process Industries.* Instrument Society of America, Research Triangle Park, NC.
Harris, C. J., and S. A. Billings, editors (1985). *Self-Tuning and Adaptive Control: Theory and Applications.* Peter Peregrinus Ltd., London.
Kuo, B. C. (1982). *Automatic Control Systems,* Prentice-Hall, Inc., Englewood Cliffs, NJ.
Landau, Yoab (1979). *Adaptive Control.* Marcel Dekker, Inc., New York.
Leitmann, George (1986). *The Calculus of Variations and Optimal Control.* Plenum Press, New York.
Levine, Willan, editor (1996). *The Control Handbook.* CRC Press LLC, Boca Raton, FL.
Liptak, Bela, editor (1999). *Instrument Engineers? Handbook—Process Control.* CRC Press LLC, Boca Raton, FL.
Major, Michael (1998). "Model-Free Adaptive Control on an Evaporator," in *Control.* Putman Publishing Company, Chicago.
McMillan, G. K. (1989). *Continuous Control Techniques for Distributed Control Systems.* Instrument Society of America, Research Triangle Park, NC.
McMillan, G. K., and C. M. Toarmina (1995). *Advanced Temperature Control.* Instrument Society of America, Research Triangle Park, NC.
Miller, Thomas, III, Richard Sutton, and Paul Werbos, editors (1990). *Neural Networks for Control.* MIT Press, Cambridge, MA.
Morari, Manfred, and Evanghelos Zafiriou (1989). *Robust Process Control,* Prentice-Hall, Inc., New York.
Narendra, Kumpati, and Anuradha Annaswamy (1989). *Stable Adaptive Systems.* Prentice-Hall, Inc., New York.
Sastry, Shankar, and Marc Bodson (1989). *Adaptive Control, Stability, Convergence, and Robustness.* Prentice-Hall, Inc., New York.
Shinskey, F. G. (1981). *Controlling Multivariable Processes.* Instrument Society of America, Research Triangle Park, NC.
White, D. A., and D. A. Sofge, editors (1992). *Handbook of Intelligent Control: Neural, Fuzzy, and Adaptive Approaches.* Van Nostrand Reinhold, New York.
Widrow, Bernard, and Samuel Stearns (1985). *Adaptive Signal Processing,* Prentice-Hall, Inc., New York.

5

EXPERT-BASED ADAPTIVE CONTROL: CONTROLSOFT'S INTUNE ADAPTIVE AND DIAGNOSTIC SOFTWARE

Tien-Li Chia and Irving Lefkowitz

Tuning control loops has been a problem since the earliest applications of PID control. It has been noted that a significant fraction of PID control loops in plant operations are not properly tuned because of a lack of manpower, a lack of knowledge for tuning, and frequent changes of process with changes of time, raw materials, or environment settings. A mistuned controller may not be detected simply because the process is running in steady state. Unfortunately, no comprehensive statistics exist on the economic losses resulting from such improperly tuned loops.

In a typical scenario in plant operation, the occurrence of a disturbance or change of setpoint in an improperly tuned loop causes the process to exhibit an unsatisfactory transient response. As a result, the operator often shifts the loop to manual control and, after the process has returned to its steady state, shifts the loop back to automatic control. This practice greatly reduces the benefits of automatic control. It also increases the burden on the operator and results in off-spec production.

The benefits of well-behaved control loops are clear; therefore, much effort has been put into improving loop tuning techniques. A number of effective tuning methods have been developed over the past several decades. These include the Ziegler–Nichols method, the reaction curve method, the Åström relay method, and the internal model-based control method. These methods have led to the development of many devices and software packages for loop tuning that have contributed significantly to improving the performance of processes running under automatic control.

ON-DEMAND TUNING AND ADAPTIVE TUNING

On-demand tuning is a tuning process initiated by the user. Typically, a test signal is generated and injected into the process that is being tuned. A step change in control output is often used as a test signal. The process response to the step change in control output is recorded and analyzed. Controller parameters corresponding to desired loop behavior are derived based on the observed response. Although effective, on-demand tuning suffers the disadvantages of requiring:

1. Continual attention on the part of the user to determine when the loop needs tuning
2. The user's active participation in perturbing the process and then waiting for the response

The latter can be time consuming, particularly for slow processes where several hours may be required to collect enough data for satisfactory loop tuning. Connected with this is still another limitation: often production people do not want their process perturbed, by anyone, at any time.

Therefore, loop tuning remains a problem in many production plants because of the following three factors:

1. *Insufficient manpower.* Loop tuning is a time-consuming process and, with the situation of insufficient engineering manpower in many plants, loop tuning is often a low-priority issue.
2. *Frequent changes in operating environment.* In some applications, the dynamic behavior of the controlled process varies over time because:
   (a) The process is nonlinear and the operating conditions are changing (e.g., the process throughput rate is changing), or
   (b) the plant parameters are changing with time (e.g., change of process gain due to gradual fouling of a heat transfer surface). Under such circumstances, one set of tuning constants cannot satisfy the entire operating range.
3. *Perturbation signal.* For some sensitive processes, it is not acceptable to perturb the process for the purpose of tuning.

The concept of adaptive control/automatic tuning consists of a control system that can change controller parameters automatically to adapt to changes in environment or operating conditions, without any human interference. The idea of adaptive control is attractive. Not only is it able to deal with time-varying processes,

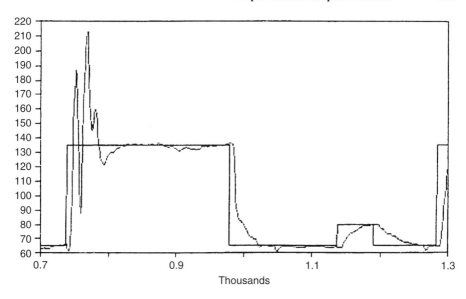

Figure 5.1 System gain changes at different operating ranges for a laboratory heat exchanger.

but also manpower shortage is no longer an issue. Adaptive control without any perturbation signal is even more attractive in dealing with some sensitive or critical processes.

Although it does have its merits and advantages, not all control loops require adaptive control. For many time-invariant systems, on-demand tuning can be very effective. For some slow but known time-varying processes or nonlinear processes, a well-tuned gain schedule controller works well.

The effect of nonlinearities in the temperature response of a heat exchanger is shown in Figure 5.1. The controller works fine when the setpoint is at a low temperature value; however, when the setpoint is changed to a higher value, the control system becomes unstable.

HISTORY AND MILESTONE LITERATURE

In the application of adaptive control (also known as self-tuning control), the controller will change its own controller settings during normal closed-loop operation to adapt to changes of the process, with the objective of maintaining optimal control performance at all times. The idea of adaptive control is very appealing because of the loop tuning problems cited earlier.

The adaptive controller is designed to deal with changes in the operating environment, but it can also correct improperly tuned control loops. By tuning loops automatically without human intervention, adaptive control eliminates the manpower and time required for adaptive tuning. Also, adaptive tuning can provide correct tuning in closed-loop operation without the need for perturbation signals to be injected into the process; this is crucial to many reactors and critical processes.

PID controller tuning had its start with the seminal work of Ziegler and Nichols in 1942, followed by Cohen and Coon's refinements of the method in 1953. The critical contribution was the recognition that most processes to which PID control is applied, although characteristically described by complex, high-order dynamics, may be effectively approximated by a first-order lag plus deadtime model, and that the parameters of this model are easily determined experimentally and are convertible to useful PID controller settings. Although this approach was quite successful for manual tuning of PI and PID loops in the field, it had limitations with respect to its underlying theoretical assumptions and with respect to the quest for automatic computer implementation of controller tuning.

Research on more general state-based approaches to adaptive control can be traced back to the 1950s. In a significantly new approach to the field, K. J. Åström and B. Wittenmark proposed, in 1973, a self-tuning regulator (STR) based on optimal control theory applied to a general nth-order linear model of the system. The model coefficients are identified by a recursive least squares (RLS) method applied to observed process input and output signals. It became apparent that the performance of their STR depended directly on the properties of the RLS identification algorithm. An important follow-up contribution then was the 1975 paper of L. Ljung, T. Soderstom, and I. Gustavsson, giving the conditions for convergence of the algorithm and showing that the STR satisfied these conditions.

These studies are only representative of a large outpouring of research in the field of adaptive control. A good survey paper annotating these developments was done by B. Wittenmark in 1975. Finally, we cite for general reference three fairly comprehensive books relevant to adaptive control: *Adaptive Filtering, Prediction, and Control*, by G. C. Goodwin and K. S. Sin (1984); *Theory and Practice of Recursive Identification* by L. Ljung and T. Soderstrom (1983); and *Automatic Tuning of PID Controllers* by K. J. Åström and T. Hagglund (1988).

While most of the academic world followed the footsteps and the framework of the mathematically rigorous approach originally proposed by Åström and Wittenmark, some other researchers, such as J. Pollard and C. Brosilow at Case Western Reserve University (Pollard, 1985), and B. Bakshi and G. Stephanopolos at the Massachusetts Institute of Technology (Bakshi, 1992), took a different approach to solving the adaptive control problem. They chose to use heuristic and pattern recognition-based techniques. The heuristic approach was also adopted by industry with good success through the Foxboro Exact controller and the ControlSoft INTUNE product (Chia, 1992).

A neural net approach was also used, more recently, for developing adaptive control solutions (Sobacic et al., 1991). In this approach, the neural net characterizes the system's input/output behavior through accumulated input/output data. The resulting model maps the cause-and-effect relationships of the system, which can be used in problem diagnosis, statistical quality control, and/or process control.

To distinguish between the two different approaches in adaptive control, we will refer to adaptive control similar to the (STR) approach proposed by Åström and Wittenmark as identification-based adaptive control (IAC). We will refer to the heuristic and pattern recognition-based approach as expert-based adaptive control (EAC).

ADAPTIVE CONTROL STRUCTURE AND UNDERLYING PRINCIPLES

The structure of an adaptive controller consists of two basic elements: the identifier block and the adaptive block (Figure 5.2). The inputs of the identifier block include such process information as the current values of the process variable (PV), control output (CO), and setpoint (SP), auto/manual mode, etc. The identifier block identifies and characterizes the behavior of the process, given the process information. The output of the identifier block is then passed to the adaptive block. The function of the adaptive block is to come up with a control decision based on the output of the identifier. If a controller is part of the adaptive block, the control output is calculated and then sent to the system directly. In this case, it is an adaptive controller.

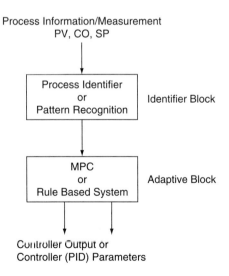

Figure 5.2 Adaptive control structure.

Alternatively, the adaptive block may also calculate and update the parameters of the controller that is running in closed loop. In this case, it is an adaptive tuner rather than an adaptive controller (Figure 5.3). In this chapter, we will not distinguish these two cases because it does not affect the overall behavior of the adaptive control.

In the identification approach, the underlying principle is that the process model can be identified in closed loop and then control can be calculated based on the identified process model. The function of the identifier block is to identify the dynamics of the system that is being controlled. Usually it is a statistical identification scheme, which can be a least squares identifier, a maximum likelihood identifier, or a nonlinear identifier. Once the model of the process is available, controller parameters (tuning parameters) can be derived using one of the many model predictive control (MPC) schemes. Alternatively, if a proportional-integral-derivative (PID) controller is used instead of MPC, an equivalent set of PID parameters can be derived so that the PID controller approximates the MPC (Rivera et al., 1986). The identification and controller calculations are carried out recursively at each sample period.

The self-tuning regulator proposed by Åström and Wittenmark originally consisted of a least squares identifier with a minimum variance controller. The STR system has a nice property: it will converge to the optimal controller even if the noise is not white (in which case, the system dynamics cannot be properly identified by a least squares algorithm; see later sections). The performance of this class of adaptive control hinges heavily on the identification scheme; hence, we label this class as identification-based adaptive control.

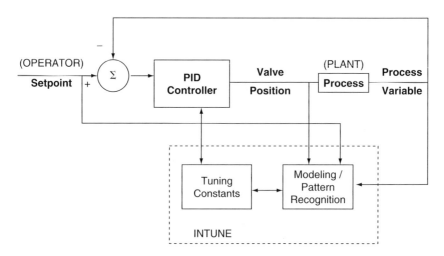

Figure 5.3 *INTUNE in operation (PID mode).*

In the expert-based adaptive control approach, the identifier is a pattern recognition engine that characterizes the transient behavior of the closed-loop response based on the observed controller output and process variable signals. The adaptive block is usually a rule-based expert system. Once the transient behavior is identified and validated, the behavior information is transformed into a type of information that can be understood by the adaptive block, such as a damping ratio, a percent of overshoot, or a settling time. The adaptive block will use this information to determine how to change the controller parameters so that the desired closed-loop response can be achieved and maintained.

In the EAC approach, little or no emphasis is placed on statistics-based process identification algorithms. A more direct approach is taken—instead of identifying the process model, it looks at the closed-loop response directly. The set of heuristic rules, which dictates how the controller parameters should be adjusted, differs from one system to another. Some systems use the Ziegler–Nichols tuning method, while others use tuning rules that mimic how an experienced control engineer would tune a loop online. In the expert-based adaptive control approach, the development of the pattern recognition module and the expert module requires more real-life control expertise. In particular, the performance of the system depends on the developer's experience with loop tuning and his or her understanding of feedback and the underlying control theory. One of the key challenges is to distinguish the effects of disturbance dynamics from the process input/output dynamics so that the adaptive controller does not produce incorrect tuning results because of disturbance dynamics.

IDENTIFICATION-BASED ADAPTIVE CONTROL

Process Model

In identification-based adaptive control an explicit model structure is assumed. A commonly used model structure is the autoregressive moving average model:

$$y(t) = -a_1 y(t-1) - a_2 y(t-2) - \cdots - a_{na} y(t-na) \\ + b_0 u(t-d) + b_1 u(t-d-1) + \cdots + b_{nb-1} u(t-d-nb+1) \\ + e(t) + c_1 e(t-1) + \cdots + c_{nc} e(t-nc)$$

where

$y(t)$ is the value of the output of the system at time t
$u(t)$ is the value of the input of the system at time t
$e(t)$ is the noise
d is the deadtime of the process expressed as the number of sample times.

Note that if the noise input can be assumed to be white, then $c_1 = c_2 = \cdots = c_{nc} = 0$.

The preceding model can be expressed more compactly in the form

$$A(q^{-1})y(t) = q^{-d}B(q^{-1})u(t) + C(q^{-1})e(t)$$

where

$$A(q^{-1}) = 1 + a_1 q^{-1} + a_2 q^{-2} + \ldots + a_{na} q^{-na}$$

$$B(q^{-1}) = b_0 + b_1 q^{-1} + b_2 q^{-2} + \ldots + b_{nb-1} q^{-nb+1}$$

$$C(q^{-1}) = 1 + c_1 q^{-1} + c_2 q^{-2} + \ldots + c_{nc} q^{-nc}$$

and where q^{-1} denotes the unit delay operator, that is, $q^{-1}y(t) = y(t-1)$, $q^{-2}y(t) = y(t-2)$, etc.

Model Identification (Identifier Block)

The system dynamics can be written as (in the case of LS formulation):

$$y(t) = \phi^T(t)\theta + e(t)$$

where

$$\phi^T(t) = [-y(t-1), \cdots -y(t-na), u(t-d), \cdots u(t-d-nb+1)]$$

$$\theta^T = [a_1, \cdots a_{na}, b_0, \cdots b_{nb-1}]$$

Any recursive identification algorithm in the form of

$$\hat{\theta}(t+1) = \hat{\theta}(t) + k(t+1) * [y(t+1) - \phi^T(t+1)\hat{\theta}(t)]$$

where

$$\hat{\theta}(t) = \text{estimate of the vector } \theta \text{ at time } t$$

can be used for identification purposes as long as the gain matrix $k(t+1)$ is selected so as to ensure convergence of the recursive algorithm. One obvious choice is the recursive least squares algorithm (RLS) (recall the nice property of STR: even though the LS may generate bias estimates in the case of nonwhite noise, the controller is still optimum).

Derivation of the recursive least squares formula can be found in many textbooks. The result is given here for convenience:

$$k(t+1) = P(t+1)\phi(t+1)$$

$$P(t+1) = P(t)\left[I - \frac{\phi(t+1)\phi(t+1)^T P(t)}{\lambda(t+1) + \phi(t+1)^T P(t)\phi(t+1)}\right]\Big/ \lambda(t+1)$$

where λ = time-varying forgetting factor, $0 < \lambda \leq 1$.

When the assumption of white noise is not satisfactory (i.e., if any of the coefficients are nonzero), then the recursive extended least squares (RELS) may be considered:

$$\varepsilon(t) = y(t) - \phi^T(e)\hat{\theta}(t-1)$$

where

$$\phi(t) = [-y(t-1), \cdots, y(t-n_a), u(t-d), \cdots, u(t-d-nb+1), \varepsilon(t-1)]^T$$

$$\hat{\theta}(t) = [\hat{a}_1, \cdots \hat{a}_{na}, \hat{b}_0 \cdots \hat{b}_{nb-1}, \hat{c}_1, \cdots \hat{c}_{nc}]^T$$

Alternatively, the related identification scheme, recursive maximum likelihood (RML), can be used.

RELS and RML may help in identifying process dynamics (in the colored noise case); however, the benefits for adaptive control are limited according to the author's experience.

Adaptive Block

The function of the adaptive block is to generate the control signal based on the identified model parameters; in the case of an adaptive PID tuner, optimal PID parameters are generated based on the identified model parameters. The adaptive block may determine the control output signal by solving a time-based optimization problem.

The following generalized control objective can be used for the control calculation in the single-input, single-output case:

$$\text{Min} \sum_{i=0}^{h} [\alpha_i e^2(t+d+i) + \beta_i u^2(t+i)]$$

where

$$e(t) = y(t) - y_{sp}(t)$$

$$y(t) = -a_1 y(t-1) - a_2 y(t-2) - \cdots - a_{na} y(t-na)$$
$$+ b_0 u(t-d) + b_1 u(t-d-1) + \cdots + b_{nb-1} u(t-d-nb+1)$$
$$+ e(t) + c_1 e(t-1) + \cdots + c_{nc} e(t-nc)$$

and $\alpha_1, \alpha_2, \cdots, \beta_1, \beta_2, \cdots$ are known weighting coefficients.

In this formulation, h is the time horizon that usually can be specified by the user. Solution of the above problem yields the sequence $u^*(t), u^*(t+1), \cdots u^*(t+h)$, which will force the sequence $y(t+d) \cdots y(t+d+h)$ to track the desired trajectory $y_{sp}(t+d) \cdots y_{sp}(t+d+h)$ that is specified by the user. The weighting factors α_i's and β_i's can be used for tuning purposes. They provide the basis for trade-off between setpoint tracking and control output variations. If setpoint tracking is the main objective, α_i's can be used exclusively while β_i's are set to zero. If smooth operation is also desirable, some nonzero, β_i's should be applied as well.

In this formulation, the sequence of control actions $u^*(t), u^*(t+1), \cdots, u^*(t+h)$ is determined at every control instant; however, only the control action corresponding to current time $u^*(t)$ is implemented. The identification and control calculation will repeat at each sample time. The solution of the foregoing control problem can be found in many optimal control textbooks.

The objective function in the minimum variance controller is:

$$\min E[y(t+d) - y_{sp}(t+d)]^2$$

where

$$A(q^{-1}) y(t+d) = B(q^{-1}) u(t) + e(t+d)$$

In MVC, the system is trying to reach a desired setpoint in the shortest time possible (next sample period) regardless of the costs associated with the control outputs. The system that is being identified can be written

$$\hat{y}(t+d) + a_1 \hat{y}(t+d-1) + \cdots = b_0 u(t) + b_1 u(t-1) + \cdots + \varepsilon(t+d)$$

$u^*(t)$ can be calculated by the formula:

$$u^*(t) = \frac{1}{b_0} \{ y_{sp}(t+d) + a_1 \hat{y}(t+d-1) + \cdots - b_1 u(t-1) \cdots - \varepsilon(t+d) \}$$

so that $y(t+d)$ will track the $y_{sp}(t+d)$.

Again, the MVC solution can be found in many articles and textbooks; it is given here for convenience:

$$u^*(t) = \frac{-G(q^{-1})}{B(q^{-1})F(q^{-1})} y(t)$$

$$C(q^{-1}) = A(q^{-1})F(q^{-1}) + q^{-d}G(q^{-1})$$

$$F(q^{-1}) = 1 + f_1 q^{-1} + \cdots + f_{d-1} q^{-(d-1)}$$

$$G(q^{-1}) = g_0 + g_1 q^{-1} + \cdots + g_{na-1} q^{-(na-1)}$$

From this solution formulation, it is clear that the deadtime estimation is crucial. If the deadtime is underestimated, the identified b_0 term will be very small (or even zero); as a result, the control $u(t)$ can be unstable. Thus, it is better to overestimate the deadtime than to underestimate it. Another issue about the MVC is that it tends to be too aggressive. According to the objective function, the control will make the system output $y(t)$ follow the setpoint in one sample period for any deviation that may be caused by disturbance or change of setpoint. This objective is too aggressive for most industrial applications. Therefore, the generalized cost function, using weighting factors α_i's and β_i's on the sequence of outputs $y(t), \cdots y(t+d)$ and control efforts $u(t), \cdots u(t+d)$, is often used.

Challenges of Identification-Based Adaptive Control

The concept of combining a recursive identification scheme with an optimal control algorithm to produce an adaptive control system sounds very appealing. However, there are some major difficulties in this approach: the majority of challenges and difficulties lie in the identification portion of the algorithm. Here we will list some of the challenges one may encounter, and present techniques that can be used to deal with the challenges.

Deadtime

The importance of the deadtime estimation is clear from the previous section. Deadtime is a nonlinear element with respect to process identification. Linear identification methods cannot estimate the process dynamics and deadtime simultaneously. Three methods commonly used to identify the deadtime are:

1. Identify multiple process models, each with a different assumed value for the deadtime, then select the model that fits the data best. The selection of the best model can be done by comparing the sum of squares of residues $\Sigma \varepsilon^2(t)$ and checking the correlation coefficients of the sequences of $\varepsilon(t)$'s. The validation of the identified model can be found in *Time Series Analysis, Forecasting and Control*, by G. E. P. Box and G. M. Jenkins (1970).
2. Employ a higher-order model and then, based on the significance of the identified parameters, estimate the process deadtime. The determination of

significance of identified parameters can be done heuristically, or it can be done by checking residuals as described in method 1.
3. Use a nonlinear estimator that could identify both the parameters of the $A(q^{-1})$, $B(q^{-1})$, and $C(q^{-1})$ matrices and the deadtime simultaneously.

Startup and PRBS

During the startup of adaptive control, the control response may be very poor until a reasonable model has been identified. To speed up the identification process, a pseudo-random binary signal (PRBS) is often suggested and used during the startup stage. PRBS is designed to excite the system for the purpose of better and faster identification; however, external excitation is not acceptable for many processes. Alternatives include using a less disturbing step test, or starting with a preliminary model. Although this latter alternative may ease the startup problem, it diminishes the usefulness of adaptive control at the same time, since human interference or some process knowledge is required to generate the preliminary model.

Model Order

The selection of model order is an important task in adaptive control. In principle, the more accurate the model, the better the output predictions. This seems to suggest the selection of a higher-order model. However, the higher the model order is, the more parameters need to be identified, which increases the identification task. More importantly, assuming the case of a perfect model and no constraints on the control outputs, one can control the system such that the process variable will track the setpoint exactly, except trailing by the deadtime (Figure 5.4). Even though perfect control can be achieved, it will not be implemented in most real-life applications

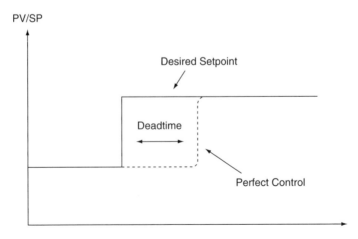

Figure 5.4   Perfect control can be achieved for a perfect model.

because of the risks of modeling errors and uncertainty of environmental influences (as is pointed out in the case of MVC). Hence, the benefits of an accurate model diminish drastically. This point is made even clearer in the case of an adaptive tuner applied to a PID controller. No matter how many model parameters are identified, they all eventually map into a three-term PID controller (Figure 5.5), while an equivalent first-order system (a first-order system is defined by three parameters, namely, model gain, time constant, and deadtime) can be used instead. In fact, most chemical processes can be represented adequately by a first- or second-order system, and in practice, the lower-order model works well and does not unduly complicate the identification issue.

Noise Level

All identification methods are sensitive to noise to a certain degree and will produce incorrect results when the noise exceeds a certain level. The LS method is known to produce biased parameter estimates in an ARMA model when the data is corrupted by measurement noise. (According to the author's experience, the LS method breaks down when the noise-to-signal ratio exceeds 0.2.)

The instrumental variable method is a variation of LS that is more resistant to noise, but at the expense of requiring more computation. In this method, essentially, the measured process output variable, $y(t)$, is replaced by $y_M(t)$, a model-generated

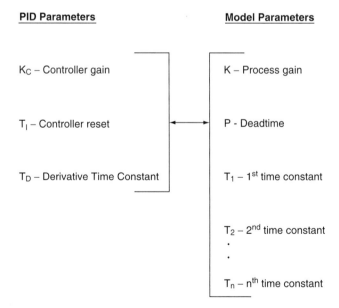

Figure 5.5 Mapping between model parameters and PID parameters.

variable based on the actual system inputs and the current estimates of the model parameters. Thus, using the notation of the section on the process model,

$$y_M(t) = \phi_M^T(t)\hat{\theta}$$

where

$$\phi_M^T(t) = [-y_M(t-1), \cdots -y_M(t-na), u(t-d), \cdots u(t-d-nb+1)]$$

$$\hat{\theta}^T = [\hat{a}_1, \cdots \hat{a}_{na}, \hat{b}_0, \cdots \hat{b}_{nb-1}]$$

The vector $\phi_M(t)$ is termed the *instrumental variable* vector; it has the property of being uncorrelated with the noise input to the system, hence reducing the bias inherent in the basic LS method. The parameter vector $\hat{\theta}$ denotes the set of system parameters determined in the most recent iteration of the identification algorithm.

Identifiability

The assumption of identification-based adaptive control is that the model can be identified during closed-loop operation. A necessary condition for closed-loop identifiability is that the process input signal be independent of the process output signal. This will obviously not be the case in the closed loop, that is, where the input is determined by the system feedback. The degree of independence between input and output may be characterized by their cross-correlation function or by the degree of persistent excitation; the greater the degree of independence (or randomness), the better the identification results will be (this is why PRBS and arbitrarily timed setpoint changes have been proposed and used). In adaptive control, the process input signal (controller output) is calculated as a function of past process outputs. This means that the persistent excitation requirement will not be satisfied; therefore, the accuracy of the closed-loop identification is often questionable. This presents a major problem for identification-based adaptive control.

The challenges of identification-based adaptive control are found mostly in the identification portion of the algorithm, as described previously. Most of the challenges can be addressed with some engineering maneuvering, except for the identifiability issue. In fact, *the most serious drawback of identification-based adaptive control is the fundamental conflict between the needs of process identification and the objectives of control in closed-loop operation.* For model identification, persistent excitation is required to guarantee good results; therefore, a perturbation signal is often required (e.g., a step change of setpoint or injection of a PRBS signal). For closed-loop control operation, on the other hand, smooth process behavior is desired—frequent setpoint changes or a perturbing signal often cannot be tolerated. This conflict undermines the practical usefulness and, hence, the chance for success for most identification-based adaptive control applications. This fundamental problem was pointed out in the author's article "Some Basic Approaches for Self-Tuning Controllers" (Chia, 1992).

In order to safeguard against the problem just cited, the control system needs to include a safeguard device that works on top of the adaptive control scheme to prevent failure of the identifier. Typically, the safeguard device is a rule-based, heuristic expert system that screens out ill-conditioned data. Obviously, the more complete the safeguard device, the more the system will look like an expert-based adaptive controller possessing both the advantages and the limitations of the expert-based system.

EXPERT-BASED ADAPTIVE CONTROL: CONTROLSOFT'S INTUNE

INTUNE Product History and Evolution

ControlSoft's INTUNE adaptive control was first developed in 1986 using a hybrid approach (i.e., a combination of model-based and expert-based techniques). A pattern recognition-based expert system was developed to recognize key attributes of process behavior and to exclude poor data from the identification algorithm. An identification scheme was used to identify the parameters of a first-order plus dead-time model, $\hat{G}_p$, characterizing the process dynamics:

$$\hat{G}_p = \frac{\hat{k} e^{-\hat{D}s}}{\hat{T}s + 1}$$

where

$\hat{k}$ = estimated process gain
$\hat{D}$ = estimated deadtime
$\hat{T}$ = estimated lag time constant
s = Laplace operator

An optimally tuned internal model-based control (IMC) is developed for the process (Figure 5.6). Then the parameters of an equivalent PID control are determined based on the IMC-based tuning results. For details of internal model-based control and tuning techniques, refer to Rivera et al. (1986).

In the hybrid INTUNE development, we experienced all the challenges of the identification-based adaptive control (IAC) approach mentioned in the previous section. The results of the identification scheme are often questionable because the persistent excitation condition is not satisfied. As explained in the previous section, a safety device is needed to make the identification function properly. Therefore, major

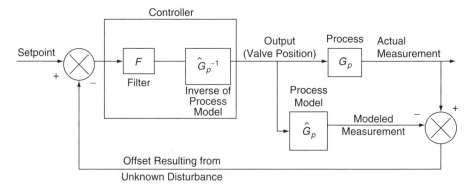

Figure 5.6 Internal model-based controller.

efforts were spent on developing the pattern recognition feature in order to filter the useful information. Once the behavior of the closed loop system was captured, verified, and understood by the pattern recognition module, we learned a lot about the system and its dynamics; as a result, the value of the identification scheme diminished. Even the controller parameter calculations became unnecessary since the tuning decision can be made based on the observed transient behavior. Even though model identification may not be necessary for controller tuning purposes, it still has value for modeling and simulation purposes.

Hence, the INTUNE system has evolved from a hybrid system to a pure expert-based adaptive control system that uses pattern recognition techniques to identify the process behavior and heuristic tuning rules to determine new controller settings. INTUNE monitors the process variable, control output, and setpoint in real time and updates (or recommends new parameter values if advisory mode is selected) the controller's PID settings to account for any changes in the behavior of the process, without disturbing the system.

In the process of recognizing the process dynamics, the software needs to distinguish those attributes characterizing the closed-loop dynamics from data contaminated by disturbances and noise. Also, the software needs to be aware of the state of the controller, such as control output saturation, manual/auto status, and change of setpoints. These all need to be monitored and taken into account when making tuning decisions. Therefore, the pattern recognition portion of the software needs to be able to comprehend the process behavior thoroughly, as an engineer would. Once this information is available, the software should have fairly good knowledge of the process loop, allowing many engineering decisions to be made.

In order to maintain a process, many questions need to be answered, such as: Is the dynamic behavior satisfactory in closed-loop operation? How often does a critical disturbance occur? What is the size and frequency of the disturbance? Is the valve sized correctly to handle the disturbance? All of these questions should be answerable by a review of the information that is normally collected by the pattern recognition scheme.

INTUNE collects raw process data and translates it into valuable process information that is useful for engineers. This information is organized in a report that allows the user to identify problematic loops. Since the monitoring process has no direct effect on the loops' behavior, the software can monitor a large number of loops simultaneously. Thus, the software can assist process engineers in maintaining the process and in diagnosing trouble loops.

INTUNE is not a historical data logging software; rather, it collects raw process data and then intelligently converts it into information that can be used to make evaluations and corrective decisions, as an experienced engineer would do.

INTUNE's diagnostic capability goes well beyond its loop-tuning capability and, since many process control problems are not necessarily tuning related, turns out to be of tremendous value to plant personnel. Thus, this diagnostic capability can assist plant engineers maintain and improve their processes despite reductions in personnel and ever-increasing workload in the plant.

INTUNE Structure

An expert-based adaptive control system consists of two major components, similar to the IAC approach (Figure 5.1). The identifier is a pattern recognition engine that characterizes the transient behavior of the closed-loop response based on the observed controller output and process variable signals. The adaptive block is usually a rule-based expert system. The function, structure, and techniques involved in the identification blocks and adaptive blocks of the INTUNE software are discussed next.

Identifier Block and Pattern Recognition

Event Detection

The function of the identifier in expert-based adaptive control is to recognize and characterize abnormal process behavior. In the case of INTUNE, process behavior is recognized and categorized according to the events and phenomena considered most likely to generate the observed behavior. Specifically, during the pattern recognition process, INTUNE looks for specific events and patterns in the process, recognizing both operator-induced events and process events. Operator events are those actions initiated by the operator that cause the control system to react, such as setpoint changes, PID setting changes, or switching the controller from auto to manual.

Process events are recognizable process behaviors that may be induced by operator events or results of disturbances caused by upstream processes.

Recognizable process events include peaks in the CO and PV signals, CO saturation, PV oscillation, sluggish PV response, and rates of change of PV and CO signals. Operator-induced events are monitored and recorded separately because the data will be used differently. This is because response data following operator-induced events should characterize the true process dynamics. Some of the operator-induced events monitored by INTUNE are:

- *Setpoint change.* SP is changed significantly relative to the PV dead band.
- *Manual/auto mode.* Controller mode is changed; adaptive tuning is inactive when in manual mode.
- *PID change.* User manually changes the PID settings
- *Startup.* Tuning software is first brought online and the PV deviates from the SP.

After an operator event has been detected, INTUNE will attempt to make PID parameter adjustments if data containing sufficient information for decision making was collected (see later sections).

Pattern Recognition

INTUNE monitors certain characteristics of the process behavior to determine when a process event has occurred. A process event is a specific pattern of behavior that fits certain characteristics that INTUNE is looking for. The measures that are used to describe the process behavior are PV dead band, response time, current CO zone status (Figure 5.7), current PV zone status (Figure 5.8), respective times that PV and CO data stay in a particular zone, and the rate of change (slope) of the PV and CO signals. The CO and PV responses are monitored separately. The decisions on adaptive control and loop diagnostics are done by referencing both PV and CO behaviors.

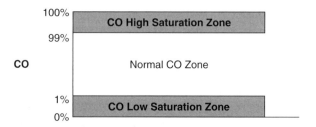

Figure 5.7   Graph of CO data range showing the CO zones.

Expert-Based Adaptive Control   221

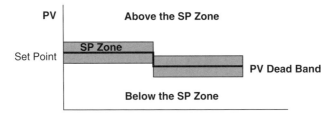

Figure 5.8 *Graph of PV data range showing the PV zones.*

By keeping track of the foregoing information on the CO and PV signals, process behavior can be described qualitatively without saving the complete data set. In fact, INTUNE can almost depict the PV trend chart using the collected characteristic information.

Since the process is continuous, data is monitored and collected continuously. Hence, it is necessary to separate the data into segments, each representative of a particular transient process behavior, and make decisions based on these observed behaviors. A behavior that can be used to mark the end of a transient response is when the system reaches a steady state and the PV is tracking the setpoint. However, there are many other criteria that can be used to mark an end-of-process event, such as the occurrence of an operator-initiated event, the observation that the PV is outside the noise band for too long a period or that the CO is saturating, or detection of a strong PV oscillation. Once the end-of-process event is confirmed, the software attempts to validate the event. Then the event occurrence with its associated statistics is stored and the controller parameters are adjusted as necessary.

Once the end of an event has been confirmed, INTUNE examines the recorded information to verify that the process event is valid. The validation is done by checking whether the observed oscillations of PV and CO are related (i.e., by referencing their respective transients), and by checking the PV data for slope, height and duration of peaks, decay ratio, and the speed of traversing the SP zone (Figure 5.9).

All of the peak characteristic information, such as the peak height ratio, peak span ratio, and time of passing through the SP zone, is normalized. These characteristics are checked against threshold values. Threshold values are selected heuristically, which helps to validate the related peaks and to reject the unrelated peaks that may have been caused by disturbances. This ensures that the recorded peaks represent the dynamics of the process. If an event is not validated, its information will not be used for adjusting the PID values.

222 Techniques for Adaptive Control

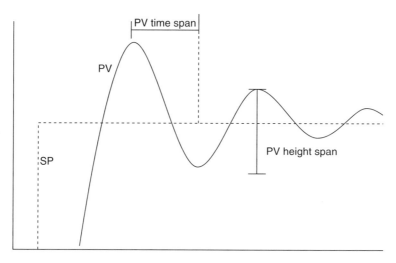

Figure 5.9 PV, SP graph showing peak time span and peak height span.

Figure 5.10 shows a typical sequence of events with INTUNE in operation. The numbers in the figure identify the following events:

1. The system recognizes the instability and the controller parameters are adjusted accordingly.
2. Setpoint changes are detected.
3. Information is collected and recorded. INTUNE recognizes that this PV deviation is caused by a disturbance since the peaks are not related to the peaks observed in event 2. This conclusion is reached by carefully checking and validating the observed behavior.
4. INTUNE observes another disturbance that has entered the system. The controller response is too aggressive. This event is verified by referring to the oscillatory response to the setpoint change in event 2.
5. The controller parameters are adjusted further and the control system behaves nicely in suppressing the disturbance, according to the user's specifications.

Adaptive Block: Rule-Based Tuning and Diagnostics
Rule-Based Tuning of PID Parameters

Once a process event has been identified and verified, INTUNE will use the information to make PID tuning adjustments. Two layers of PID adjustment are provided—basic tuning and advanced tuning. The tuning is done heuristically, just as an experienced engineer would do when tuning loops online. The system will also try to identify the dynamics of the system whenever information-rich data is available.

Expert-Based Adaptive Control 223

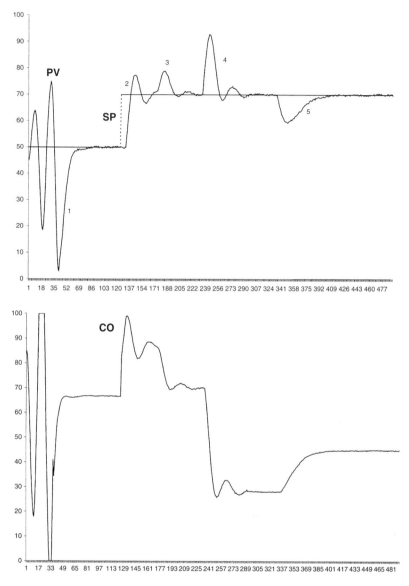

Figure 5.10 INTUNE in operation.

The identified model is verified by checking residues. If the level of confidence is high on the identified model, the model will be presented to the user. However, the identification portion is not necessary for loop tuning purposes, since the direct, heuristic-based tuning is more efficient and robust. This comes as no surprise, as pointed out previously in the identification-based adaptive control section.

Basic Tuning

INTUNE checks the process response speed against the user's desired response speed (slow, medium, or fast). If the observed response satisfies the specification, no tuning adjustment is done. Otherwise, the PID settings are adjusted based on the observed response and the current PID settings.

If the user selected a slow response, INTUNE will try to adjust the PID settings to produce minimum to zero overshoot in the closed-loop response. For a medium response there is an overshoot of 10% to 20%. If a fast response is selected, there is an overshoot of no more than 35%. The PID settings are modified depending on the observed process behavior and the current PID settings. Following are some process events that may cause INTUNE to adjust the tuning parameters:

- *Well-behaved response*. A well-behaved response has a decay ratio that is within the user-specified desired response range. In this case, no adjustment of the PID parameters is necessary.
- *Sluggish response*. A sluggish response is detected when the PV is nonoscillatory and deviates from the setpoint for too long. INTUNE will increase the controller speed according to the observed PV speed.
- *Oscillatory response*. When an oscillatory response is confirmed, the overshoot and the decay ratio will be checked against the user's specified response. PID settings are adjusted accordingly.
- *CO saturation*. A CO saturation event is detected when the PV is oscillatory and the CO enters either the high saturation zone or the low saturation zone. The controller parameters are adjusted accordingly.

Advanced Tuning

Since the qualitative behavior of the process is known to the system, INTUNE can, with some simple calculations and validations, recognize certain patterns in the process behavior that are caused by inappropriate PID settings, just as an experienced engineer would do. Many bad behaviors caused by incorrect PID settings can be recognized by checking their characteristics. Such characteristics as too-strong proportional action, too-weak integral action, or inadequate derivative action result in distinct and recognizable transient behaviors. Once these behaviors are recognized and verified, the controller parameters are corrected accordingly. Some examples of how these behaviors may be characterized are as follows:

Bad PID Combinations	Characteristics of Behavior
Excess D or P	For high noise level and for nonintegrating processes, the CO can be erratic if the D or the P is too large
Strong D	In a relatively noise-free system, one-sided peaks can be caused by a strong D action
Weak I and strong P	Weak I actions cause a long tail. A combination of a weak I and a strong P causes one-sided damping while the peaks approach the SP
Strong I	A continuous oscillation with a decay ratio close to 1 can be caused by a strong I

Process Type

In both basic and advanced tuning, it is important to distinguish between the process types, since the tuning rules are different. For example, the behavior of an integrating process (such as a level control loop) is very different from that of a nonintegrating process. Thus, different sets of rules need to be developed for the different process types. It is also useful to distinguish among process loops. For example, temperature loops are usually slow and have a low noise level, while flow loops are much faster and can be quite noisy.

Safety Issues: PID Limits, Advisory Mode

INTUNE provides a couple of safety devices in conjunction with its adaptive tuning. The first one is PID limits; it allows the user to enter limits on the P, I, and D terms independently. INTUNE checks against these limits before adjusting the PID settings. This allows the user to control the pace and the range of PID tuning. However, there is a downside to this. The control loop performance may be potentially limited if the PID limits are set improperly.

The second safety device is the mode of operation. INTUNE can operate in either adaptive mode or advisory mode. In advisory mode, INTUNE prompts the user with the recommended tuning parameter changes instead of making the changes automatically. Thus, the advisory mode is designed to give the user more control of the tuning process, that is, the downloading and change of controller parameters are implemented only if the user instructs the software to do so.

INTUNE Loop Diagnostics

In order to properly maintain a process control system, maximize its performance, and do some preventive maintenance, detailed knowledge of the process and its behavior is needed. Operators and plant engineers responsible for daily operation of the system usually have intimate knowledge of the process behavior; however, it is hard to quantify this process knowledge and to transfer it easily to process engineers and plant managers. Often a perception of how well the process is behaving may be developed by periodically checking the operation screen. However, this perception can often be deceiving, particularly for slow processes. For example, the author, once asking about a temperature loop in a plant, was told it was stable and controlled well. However, when the user began monitoring the loop and logging the data, he found that the loop was actually oscillating persistently, with a swing of $20°$ and a cycle period of 2 to 3 hours.

Even though operation personnel are knowledgeable about the process, it is difficult to describe process behavior qualitatively. For example, a process engineer may need to know: How well does each process loop track its setpoint? Are there any disturbances or upsets in the process? How often? What is the duration and magnitude of the disturbance? Process engineers may also need to know whether the process was designed properly or whether the control is capable of handling the disturbances that enter the system. Also important are measures of the condition of physical components of the control loop (e.g., the magnitude of control valve hysteresis) and whether or not the sensor is calibrated and functioning properly. Such information is needed to assist the user in maintaining and improving the process.

For process analysis purposes, the user needs to have a descriptive knowledge of aberrant process behavior (such as frequent disturbance inputs and persistent PV oscillation). Also, in order to reflect the true control performance, the statistics characterizing the process data (such as tracking error) and process behavior need to be considered in conjunction with the causes of the observed behavior. For example, if mean errors and variance of errors are calculated without considering the fact that the loop has been in manual mode for some or all of the time, or that the setpoint has had several changes, these statistics may not reflect the true process characteristic. In the case of disturbances entering the system, for example, rather than knowing the error distribution statistics, the user may be more interested in the frequency, magnitude, and duration of these disturbance events so they can be traced to an upstream process with the possibility then of taking some preventive or corrective action.

Process Performance Statistics Tracked by INTUNE

In the process of monitoring the loops, INTUNE generates fairly comprehensive information on the performance of the control loops available to the user for maintenance purposes. By proper interpretation and organization, the process data is represented and interpreted in a way that is useful for plant personnel—unlike most SPC software, which collects and processes data without considering its relationship

Expert-Based Adaptive Control 227

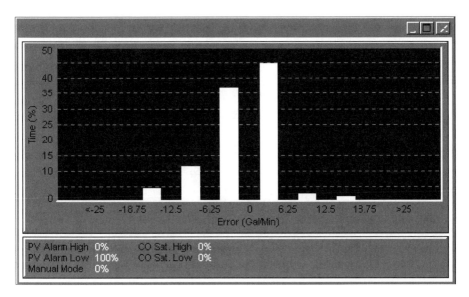

Figure 5.11 Error distribution chart.

to process behavior and operator-initiated activities. Some key statistics tracked by INTUNE to assist the diagnosis of the process loops include:

- *Error distribution chart.* The error distribution bar chart is a tool that enables the user to quickly determine the control quality of the process loop. The chart presents a histogram of the time spent by the control loop in different error regions (Figure 5.11). Each histogram bar represents the percentage of time over the period of analysis time (which is specified by user) that the error signal has been in that particular error region. A moving window database is used; this means that if the user specifies 1 week as the analysis time, the most recent 7 days of data are used for the error distribution calculation. If the deviation of the PV from the setpoint stays within a small error range, the error distribution chart will show a narrow bar in the middle of the chart. However, if the PV tends to stray away from the setpoint, then the chart will show a broader or an off-centered distribution. Therefore, browsing the error distribution chart gives the user an instant visual indication of the quality of control.
- *PV alarm/CO saturation alarm.* The PV Alarm High and the PV Alarm Low show, respectively, the percentage of time over the analysis time period that the PV values have spent in each PV alarm range. This helps the user determine whether the PV has spent too much time outside the normal operating range. The CO Saturation High values and CO Saturation Low values are also recorded. By combining PV Alarm and CO

228   Techniques for Adaptive Control

```
LOOP  2  NSTEST2                                        17-MAY-1999 14:43:09

     PROCESS VAR    50.03     CONTROL OUT   16.67     SET POINT     80.00
     PROPORTION      0.00     INTEGRAL       0.00Secs. DERIVATIVES   0.00Secs.
  TUNING PARAMETERS:
     NOISE BAND      1.00     RESPNS TIME   30.00     SAMPLE TIME   1.00
     DESIRED SPD   MEDIUM     TUNING MODE    DONE     CTRL DIRECT REVERSE

 M/T REPORT:
                   -50%  -20%  -10%    -5%    -1%    1%    5%   10%   20%   50%
  ERR DIST(%)       47    3     3      3      15    13     3    2     1    10

        P     I     D     Gain   Tau    Dlay   MCC    MPE   RSPT  SRC    TIME
 1    0.27  2.01  2.67   0.67   2.01   8.00   NA    0.22   14   DIST   MAY-17/14:21
 2    0.30  2.27  2.67   0.67   2.27   8.00   NA    0.27   15   DIST   MAY-17/14:24
 3    0.04  2.77  6.00   2.69   2.77   18.00  0.996 6.89   22   OSCI   MAY-17/14:28
 4    0.06  3.96  5.67   2.97   3.96   17.00  1.000 1.77   24   OSCI   MAY-17/14:30
 5    0.06  4.64  5.67   3.05   4.64   17.00  0.995 6.88   24   OSCI   MAY-17/14:32
 1-SCRN2 2-LOOP# 3-MODE 4-N.B. 5-RSP T 6-SMP T 7-C-SPD 8-TREND 9-D LOG 0-EXIT
    Press ">" to redraw the screen
```

(A)

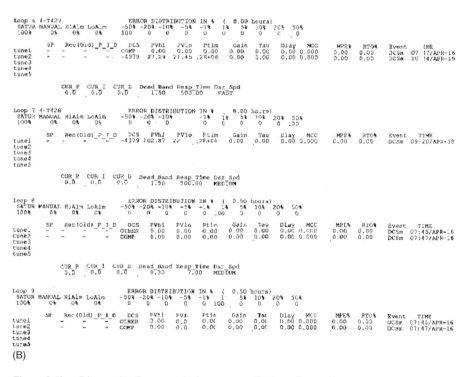

(B)

Figure 5.12   Diagnostics Report: (A) loop setup; (B) loop diagnostics.

M/T SUMMARY

1: 4-P422	2: 4-L422	3: 4-T419	4: 4-T425 sluggish	5: 4-T426 sluggish
16/07:45 DCSm	16/07:46 DCSm	16/07:45 DCSm	17/02:10 DCSm	16/07:45 DCSm
18/07:48 DCSm	16/07:48 DCSm	16/07:47 DCSm	17/19:51 OSCI	16/07:46 DTST
Event3	Event3	Event3	Event3	16/07:48 DCSm
Event4	Event4	Event4	Event4	17/02:04 OSCI
Event5	Event5	Event5	Event5	17/19:56 OSCI
M/T MODE	M/T MODE	M/T MODE	M/T MODE	M/T MODE
6: 4-T427 sluggish	7: 4-T428 sluggish	8:	9:	10:
16/07:47 DCSm	16/07:48 DCSm	16/07:45 DCSm	16/07:46 DCSm	18/07:08 SLOW
16/07:48 DCSm	Event2	16/07:47 DCSm	16/07:47 DCSm	18/07:14 SLOW
Event3	Event3	Event3	Event3	18/07:21 SLOW
Event4	Event4	Event4	Event4	18/07:33 SLOW
Event5	Event5	Event5	Event5	18/07:44 SLOW
M/T MODE	M/T MODE	M/T MODE	M/T MODE	M/T MODE

1-DIRECTORY 2-LOOP DATA 3-NEXT PAGE 4-PREV PAGE 5-RESERVED 0-EXIT/STOP
Press ")" to redraw the screen

(C) [Stop Task] ☐ Hold [Dismiss]

Figure 5.12 Diagnostics Report: (C) M/T Summary.

Saturation Alarm observations, the user can determine whether the controller is capable of handling the range of disturbances encountered, or whether the control valve is sized properly to handle the disturbance.

- *Manual Mode Analysis.* The Manual Mode Analysis chart shows the percentages of time that the controller has been in manual mode during the analysis time period. If this percentage is high, it means the user is not receiving all the benefits of automatic control and possible reasons for this should be examined.
- *Event Log.* The Event Log identifies and reports various controller operations and process behaviors that have occurred with respect to the process loop. The Event Log will log three basic types of events that affect the control of the process: (1) Operator-induced changes in the controller, such as the downloading of new PID parameters, the changing of CO or SP values, and the switching between manual and auto modes; (2) observed process dynamic behaviors, such as persistent oscillation, CO saturation, and disturbance events (including their frequencies, magnitudes, and durations); and (3) automatic INTUNE-induced tuning parameter changes, including dates and times.
- *Diagnostics Report.* The INTUNE diagnostic system monitors the process and control loops without the need for human intervention. An event report containing all the foregoing vital process information is saved in the Diagnostics Report (Figures 5.12A, 5.12B, and 5.12C). The report contains statistics and event logs for all the loops being monitored. The

report can be printed periodically (weekly or monthly) and can be used to check the system's performance over the most recent analysis time period. It can also be compared with previous reports to determine whether the process performance is improving or deteriorating.

Other diagnostic tools such as detection of valve hysteresis and sensor failure are also being incorporated in the INTUNE diagnostic capability. These are not discussed in this section since they are outside the scope of adaptive tuning.

CONCLUDING OBSERVATIONS

Each of the two approaches to adaptive control discussed here, namely identification-based and expert-based, have both advantages and disadvantages with respect to their effectiveness and reliability. These are summarized next.

Advantages and Drawbacks of Each Approach

Advantages of the Expert-Based Method

Since the expert-based approach is essentially a direct approach, that is, the controller settings are adjusted based on the observed behavior of the process rather than working through an intermediate process model, it tends to perform better and be more robust compared with the identification method. Further, the rule-based expert system involves mostly logic-type statements that generally require less computation time than the algebraic-type statements typically involved in process identification. Therefore, this approach is capable of handling faster loops.

Advantages of the Identification Approach

This method is intuitively simple and easy to implement because the mathematics of model identification and controller design can be found in many textbooks, and the method requires no particular expertise in process control and loop tuning. Thus, it is easier to develop this type of adaptive control system. It should be pointed out, however, that the designer's advantage does not necessarily transfer benefits to the user, nor does it provide any particular improvement with respect to overall system performance.

Drawbacks to the Heuristic Method

This is essentially a black box approach; hence, in general, the structure and heuristic rules used in the system are not transparent to the user. The performance of the expert-based system depends on the knowledge and experience of the developer.

Problems with the Identification Method

The major problem associated with this approach stems from the basic conflicts between identification and control, as pointed out before. As a result, the difficulty of developing an effective and working adaptive control system based on the identification approach is enormous and is underestimated by many developers.

Two-Time-Scale Aspect

An important feature of most of the adaptive control technologies discussed in this chapter—and particularly that of the INTUNE approach—is the two-time-scale nature of their implementation. In effect, process variables are controlled by standard feedback control algorithms; most often, the algorithm used is PID, but it could be internal model control (IMC), state space control, or some other direct control function. It operates in the same time scale as the process to maintain minimum deviation between process variable and setpoint, despite the occurrence of disturbances and setpoint changes. Superimposed on the direct controller is an adaptive or tuning function operating on a much slower time scale. The adaptive function adjusts the controller parameters, either periodically or as needed, so as to best satisfy the performance criteria (including such attributes as stability, speed of response, overshoot, and disturbance rejection) established for the direct control loop. The time scale of the adaptive loop is related to the mean time between significant changes in the effective dynamics of the process. These changes may be induced by process nonlinearities, time-varying parameters, large variations in operating point, constraint boundaries, etc. In the case of INTUNE, if the PID loop is performing optimally, no changes in controller parameters are necessary and, hence, no adaptive/tuning action is required.

The Real Value of the INTUNE Software

The utility of the adaptive control function provided by INTUNE is clear and demonstrable. In addition to direct sales of the INTUNE product, the technology has been licensed by many of the leading vendors of process control equipment, including Elsag Bailey, Westinghouse, Allen Bradley, Power Process Control, and Camile. All together, the number of industrial applications number in the tens of thousands, covering a wide range of processes and industries. Thus, issues of reliability, stability, and effectiveness of the underlying tuning technology have been well established through actual experiences in industry.

However, INTUNE's real value lies within its loop diagnostic capability, which goes well beyond its adaptive control capability. Capable of monitoring a large number of loops continuously, INTUNE's diagnostic features are designed to assist engineers in maintaining good process performance. Comprehensive reports on loop performance are generated periodically. By glancing at a report and comparing it with previous reports, the user can get a good sense of how well the process is behaving and can identify trouble loops. This is particularly valuable for process engineers and plant managers who need this information but do not work with the process on a day-to-day basis.

REFERENCES

Åström, K. J., and T. Hagglund (1988). *Automatic Tuning of PID Controllers*. Instrument Society of America.

Åström, K. J., and B. Wittenmark (1973). "On Self-Tuning Regulators," *Automatica* **9**, 185–199.

Bakshi, B. (1992). "Multi-Resolution Methods for Modeling, Analysis, and Control of Chemical Process Operation," Ph.D. Dissertation, Massachusetts Institute of Technology, Cambridge, MA.

Box, G. E. P., and G. M. Jenkins (1970). *Time Series Analysis, Forecasting and Control*. Holden Day, San Francisco.

Chia, T. L. (1992). "Some Basic Approaches for Self-Tuning Controllers," *Control Engineering*, December.

Cohen, G. H., and G. A. Coon (1953). "Theoretical Considerations of Retarded Control," *ASME Trans.* **75**.

Goodwin, G. C., and K. S. Sin (1984). *Adaptive Filtering, Prediction and Control*. Prentice-Hall, Englewood Cliffs, NJ.

Ljung, L., and T. Soderstrom (1983). *Theory and Practice of Recursive Identification*. MIT Press, Cambridge, MA.

Ljung, L., T. Soderstrom, and I. Gustavsson (1975). "Counterexamples to General Convergence of a Commonly Used Identification Method," *IEEE Trans.* **AC-20**, 643–652.

Pollard, J. (1985). "Adaptive Inferential Control," Ph.D. Dissertation, Case Institute of Technology, Cleveland, OH.

Rivera, D. E., M. Morari, and S. Skogestad (1986). "Internal Model Control, PID Controller Design," *Ind. Eng. Chem. Proc. Des. Dev.* **25**, 252–265.

Sobacic, D. J., Y. H. Pao, and D. T. Lee (1991). "Autonomous Adaptive Power System Control," in Proc. Third Symp. on Systems Application to Power Systems, Tokyo, Japan.

Wittenmark, B. (1975). "Stochastic Adaptive Control Methods: A Survey," *Int. J. Control* **21**, 705–730.

Ziegler, J. G., and N. B Nichols (1942). "Optimum Settings for Automatic Control," *ASME Trans.* **64**, 759.

6

KNOWLEDGESCAPE, AN OBJECT-ORIENTED REAL-TIME ADAPTIVE MODELING AND OPTIMIZATION EXPERT CONTROL SYSTEM FOR THE PROCESS INDUSTRIES

Lynn B. Hales and Kenneth S. Gritton

KnowledgeScape is the first process-control system that has integrated the powerful capabilities of real-time expert control, online adaptive and competitive neural network models, and genetic algorithm optimization into one software system designed to monitor and control any industrial process. Careful reflection on the descriptions of functions of KnowledgeScape suggests that its design can be used in a much broader sense. Specifically, KnowledgeScape can be used to embed intelligence in any process.

Typical benefits from the application of KnowledgeScape in mineral-processing applications are:

- 4–8% improvement in grinding throughput at the same or similar grind size
- 1–3% increase in flotation recovery
- 2–8% improvement in concentrate grade
- 10–40% decrease in reagent consumption

These benefits are achieved by integrating advanced, artificially intelligent tools into a single application that is focused on economic optimization.

Primary Function: Continuous Online Optimization

KnowledgeScape is an object-oriented real-time expert control system for the process industries that has built-in adaptive modeling and optimization capabilities. The primary use of KnowledgeScape is the online continuous monitoring of plant performance and the calculation of new process setpoints that maintain and optimize performance as feed and operating conditions vary.

Intelligent Software Objects

Conceptually, KnowledgeScape is designed around the idea of intelligent software objects that represent real-world plants and processing equipment. These software objects can be connected together, representing flow of material or information from one object to the next, or they can be combined with or nested within other objects, representing the concept of groups of equipment forming a circuit. These concepts are so powerful and unique that they have been patented (U.S. Patent No. 6,112,120).

Using the Past to Predict the Near-Term Future State of the System

Intelligent entities have the inherent capability to combine knowledge of past experience with the knowledge of their present state, to predict at least their near-term future state. This capability is embodied within KnowledgeScape via the use of adaptive online neural networks. Past experience is embodied within either crisp or fuzzy expert rules, as well as within the training of the online neural networks. Using past information to determine current process setpoints that will effectively drive the plant to more desirable conditions.

KnowledgeScape presents the user with a real-time expert system that allows knowledge to be recorded as both crisp and fuzzy rules. Rule-based control is centered around the concept of heuristics, which are rules of thumb that have been developed over time. These rules represent generalized experience that can be relied upon in the future when conditions are similar to those previously encountered.

Building Process Models to Predict the Future and Optimize the Process

A subtle and infrequently discussed fact is that the expert rules themselves represent an inverse model of the process. This leads to the question of what other types of models can be created for the processes and what could be done with them if they are more robust than the heuristic rules.

KnowledgeScape uses online neural network models to adaptively model the process, which adds to the unique features of the system. Instead of relying solely on past experience (i.e., the knowledge contained in the expert-system rule base), KnowledgeScape utilizes online neural networks to model more immediate process conditions. These models are trained continually, as the system monitors process conditions, and hence they contain knowledge of unique conditions that were previously unknown, as well as knowledge of more common operating conditions.

Use Predictions for Improving Performance

When neural networks are trained to provide a robust model of process performance, these models can then be queried to determine what will be the values of the predicted process parameters within the near-term event horizon. The predicted values of important process variables can be used not just by the rule-based expert system, but additionally by a genetic algorithm component of KnowledgeScape to ask the models what would be the *best* new process setpoints to achieve the desired control objectives.

This model-based optimization ability adds true artificial intelligence to KnowledgeScape in that it creates the basis for learning about the process. Rule-based expert control is inherently static concerning the process information that can be retained in the rules. In essence the extent of knowledge about the process is fixed because rules represent the knowledge of those who wrote the rules. Models, such as the online adaptive neural networks, that continually adapt to accurately represent the process as feed conditions and equipment conditions change add a learning component to KnowledgeScape. Hence, the performance of the process under KnowledgeScape control is not limited to just what the expert rules can do.

Scaling Computing Resources to Meet the Needs of the System

Crisp and fuzzy rules, adaptive models, and genetic optimization all require substantial computing resources, especially if they are online and are being used simultaneously. KnowledgeScape uses a distributed-hardware architecture that allows computers to be added at will to the control system to meet the computational needs as a system grows (Figure 6.1).

INTELLIGENT SOFTWARE OBJECTS AND THEIR USE IN KNOWLEDGESCAPE

Creating and Configuring Intelligent Nodes:

Configuring KnowledgeScape is accomplished by building the flow sheet of the plant or process being controlled using KnowledgeScape's concept of intelligent software objects. Figure 6.2A is taken from U.S. Patent No. 6,112,120, which describes the basic elements of an intelligent software object. These objects are made up of crisp

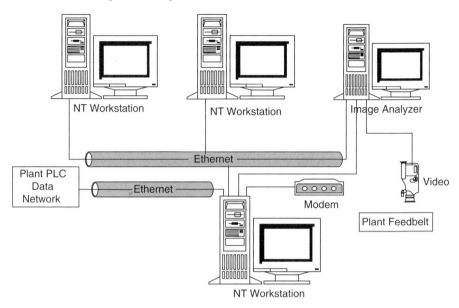

Figure 6.1   Distributed computing.

and fuzzy expert rules, models, optimizers, predictors, etc., all communicating with the necessary process sensors, and each node on a KnowledgeScape process drawing represents one ISO. Nodes can be connected, grouped together, and organized in any way that provides a meaningful representation of the process.

As sensors collect information about the process (e.g., motor speeds and temperatures), expert rules, models, predictors, and optimizers utilize this information to model the process and determine better process setpoints, and these updated setpoints are then sent back to the process via the communicator objects.

Figure 6.2B depicts the hierarchy of activities associated with configuring nodes, and all of the intelligent components of a node. Once a node is created, by dragging a "Base Node" onto the KnowledgeScape drawing canvas, any of the intelligent components may be added as noted in the figure.

Depicted in Figure 6.3 is a screen capture of the configuration window where a flow sheet is represented. Each box or node represents a piece of processing equipment. Where a node surrounds other nodes, the outer or supervisory node represents a plant area or section. Attributes are represented by the named ovals attached to nodes or connections.

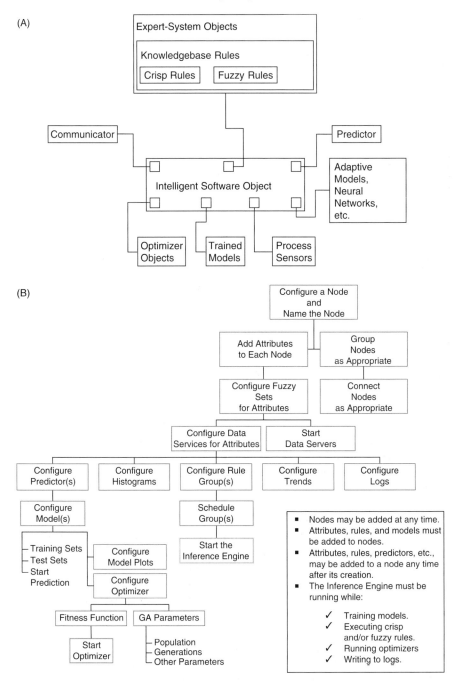

Figure 6.2 (A) Diagram of an intelligent software object (ISO); (B) Configuration of nodes and their intelligent components.

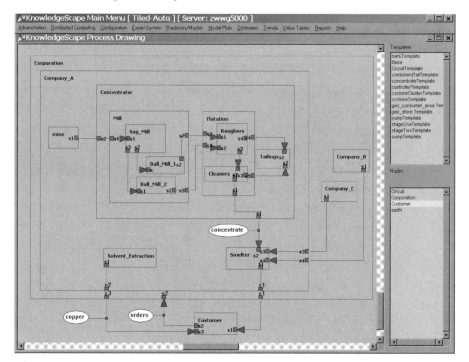

Figure 6.3 Configuration using intelligent software objects to represent a plant.

To add an attribute, the user right-clicks on the node or connection and selects the data type of the attribute. To add rules, histograms, or other components, the user right-clicks inside of a node and adds the appropriate component. To configure data servers and fuzzy sets, which are associated with attributes, the user right-clicks on the attribute and selects the desired action from the resulting drop-down menu.

Each node is a fully intelligent software object, with expert-system, modeling, optimization, and communication capabilities. Each software object optimizes its own performance, subject to constraints that are set by connected or encapsulating nodes. This results in a hierarchical configuration, where each process operation is optimized in such a way that the system works together to maximize the performance of the overall process, and not just one particular part of the plant.

The nature of the intelligence associated with each node is shown by the node menu options in Figure 6.4. By clicking on a node and looking at the menu options for that node it is seen that the options include configuring (1) a predictor or model, (2) an optimizer, and (3) process rules.

KnowledgeScape for the Process Industries 239

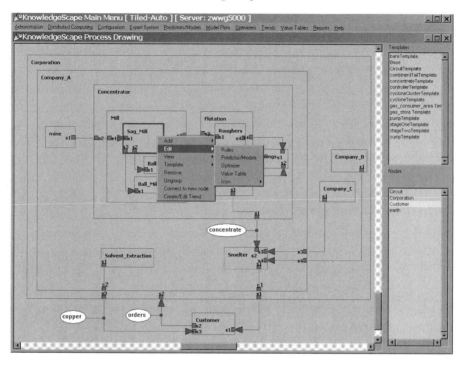

Figure 6.4   Each node, or intelligent software object has built-in artificial intelligence capabilities.

ARTIFICIAL INTELLIGENCE AND PROCESS CONTROL

Using Intelligent Nodes for Process Control

Intelligent behavior is characterized by such things as:

1. *Reasoning.* Reason and draw conclusions given an incomplete and sometimes partially wrong sets of facts.
2. *Prediction.* Given some historical information and a set of circumstances, or facts, predict future events or conditions.
3. *Generalization.* Based on experience gained in one environment or set of conditions, create generalizations that are true going from one situation to another.
4. *Abstract reasoning.* Given experience gained in one environment or from one set of conditions abstract generalizations applicable in broader terms.

KnowledgeScape has built-in artificial-intelligence tools that are integrated in a way so as to achieve behavior that can be described as intelligent, according to the criteria listed above.

In one mineral-processing application the adaptive neural network models were used with the genetic algorithm optimization functions to identify and run the plant at a combination of setpoints that were novel, and indeed unexpected. The plant manager's statement that "it taught us something we didn't already know" really sums up the value of KnowledgeScape as a tool to improve and sustain plant performance.

The point is this: *The adaptive models and intelligent components within KnowledgeScape had learned more about the cause and effect physical realities of the process than was understood by the operators or the engineers who had written the heuristic control rules contained within the expert system.*

Data I/O: Interfacing KnowledgeScape to the Real World

KnowledgeScape is primarily used as a real-time control system and thus needs to be interfaced to other systems that actually monitor and control plants. It can be thought of as a supervisory control system that monitors processes under the control of a primary stabilizing control system. KnowledgeScape calculates new process setpoints and sends them to the stabilizing control systems for implementation.

Connections via Data Servers

Data is piped into and out of KnowledgeScape by data servers associated with attributes on the process drawing. Any attribute may have a data server assigned to it, and there is no limitation on how many systems can be connected to KnowledgeScape to supply process information. Many systems have three or four of KnowledgeScape's data servers concurrently piping data into and out of KnowledgeScape.

Figure 6.5 shows the configuration screen used to configure one such data server. As the figure shows, there are many built-in drivers available within KnowledgeScape for use in communicating with a stabilizing process-control system.

Real-Time Expert Control: Expert System Rules

The Objective of the Expert Rules

The heart of KnowledgeScape is the real-time expert system that allows the designer to encapsulate knowledge about a process within both crisp and fuzzy rules. Rules and rule groups may be added to a node at any time after the node has been created. The ultimate objective of building these expert-system rule bases is to monitor process conditions and then, based on the rules, calculate new, more productive process setpoints. Other uses of the expert rule system might be the creation of smart alarm systems where alarms are monitored, characterized, and then reported in logical ways that also might point operators to quick corrective action or provide advanced analysis of causes and effects of the alarms.

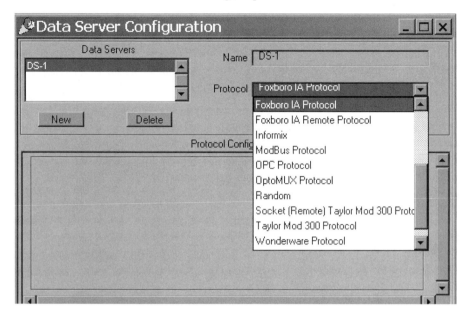

Figure 6.5 Data server configuration within KnowledgeScape.

Rule Editing: Edit Rules While Online

KnowledgeScape is designed to aid the control-system designer by compiling rules as they are entered and checking for correct syntax. Rules can also be organized according to the flows and hierarchies represented in the process drawing. This schema results in a very logical and easily maintained system.

Basic creation and editing of rules is illustrated in the accompanying screen depictions. Because of the node-based object-oriented nature of KnowledgeScape, it is possible to interact with each element of KnowledgeScape by simply pointing at it and clicking, or alternatively by going directly to the system functions via the main menu bar. This concept is shown in Figures 6.6 and 6.7.

Specifically, after configuring the process drawing with intelligent software objects or nodes, the next step is to add expert-system rules to each node to accomplish the control objectives. The expert system of KnowledgeScape is a flexible environment for creating rules that control the behavior of nodes and attributes in the process drawing. It employs a powerful rule-editing syntax within a structured execution environment.

Entering the Rule-Editing Window

Selecting the **Edit | Rules** menu option of the node's pop-up menu or selecting the Rule browser from the main menu results in the expert system rule browser opening

Techniques for Adaptive Control

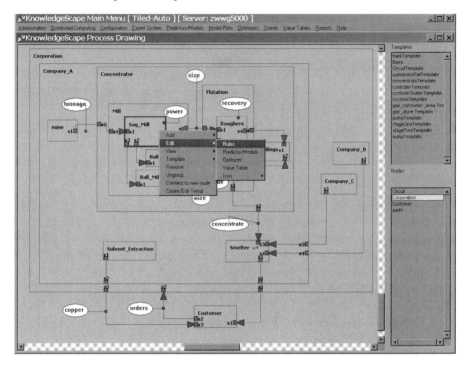

Figure 6.6 Pointing to an intelligent software object representing a SAG mill to create an associated process-control rule.

with the window title indicating the name of the node for which it was opened. The rule configuration user interface contains the node list and the rules window with tabs for each rule group and the rule schedule.

Rules are edited by double-clicking on the rule or by highlighting the rule and selecting *Edit Rule* from the right mouse-button pop-up menu. New rules are added to the group by selecting *Add Rule* from the right-click menu. Adding a rule will open the rule editor user interface with the rule text editor widget blank. Editing a rule will open the rule editor user interface with the selected rule as the text in the rule text editor widget. The user then edits the rule as KnowledgeScape checks for correct syntax. When editing is complete and the rule is parsed correctly, the rule is saved within the expert system by right-clicking on the rule and selecting one of the "accept" options from the drop-down menu. Once the rule is accepted, the current rule will be replaced or the new rule will be added to the rule group. The rule will be executed the next time the inference engine executes the group to which the rule belongs.

Building Logical Groupings of Rules

Rule groups are used to logically organize the control application by allowing the user to work at appropriate levels of detail so that complex process control issues

KnowledgeScape for the Process Industries 243

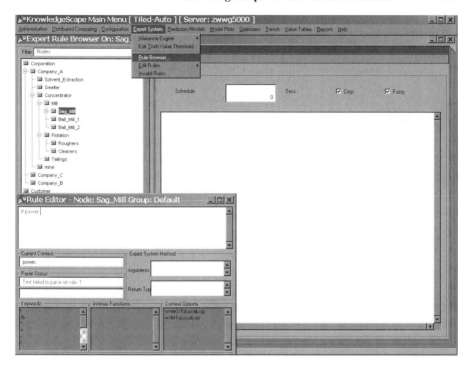

Figure 6.7 Opening the general rule editor directly from the main KnowledgeScape menu bar.

requiring hundreds of rules can be broken into manageable smaller parts. Figures 6.8 and 6.9 illustrate the concept of rule groups. Rule groups can be thought of as a bin containing expert system rules that are usually, although not necessarily, related in some way. The user decides what bin or rule group contains which rules. In other words, rule groups provide an organizational framework that helps users simplify configuration of the expert system by building a control strategy in parts.

For example, users can create a node that represents a power generation plant and define a rule group for high demand, low demand, and maintenance. One can then select the high-demand rule group and configure rules ensuring the plant operates at peak efficiency during periods of high demand. Then, the user can select the low-demand rule group and configure other rules ensuring that the plant conserves resources during periods of low demand. Later, when additional rules need to be created to ensure that environmental discharge permit limits are not exceeded during periods of high demand, users can return to the high-demand rule group and make the changes. Because the high-demand rules are encapsulated in the high-demand rule group, these changes can be made without considering the details of the rules in the low-demand or maintenance rule groups. This organizational concept is shown in Figure 6.9.

244   Techniques for Adaptive Control

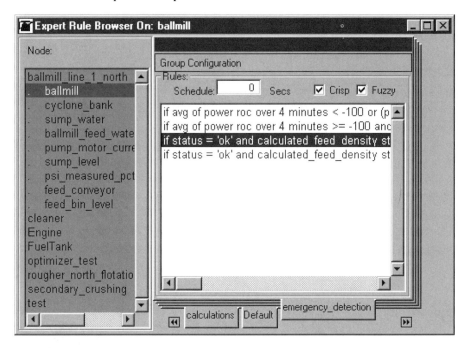

Figure 6.8   Rule groups shown as tabbed sections in the rule browser.

Figure 6.9   Rule groups for a node or intelligent software object.

Disabling Rules within Groups

Note also that rules may be temporarily or permanently disabled should the user desire to do so. To accomplish rule disabling, the user points and clicks on the rule and selects "disable" from the drop-down menu. Thereafter the rule will not be executed until a similar action reenables the rule. Disabled rules are depicted as light gray text in the rule-group window.

Controlling Rule Execution Sequences

The node-based rule browser, shown in Figure 6.10, contains a Schedule type-in box, a Fuzzy checkbox, a Crisp checkbox, and a rule list. The Schedule type-in box displays the frequency at which the expert system inference engine will execute the selected rule group. The value in the type-in box is an integer number of seconds.

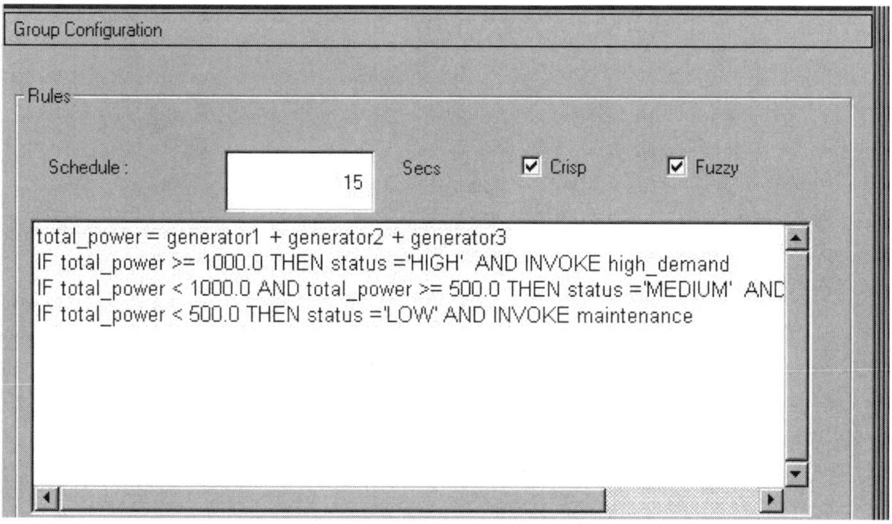

Figure 6.10 Rule group browser.

The Fuzzy checkbox controls whether rules that employ fuzzy logic are displayed in the rule list. The Crisp checkbox controls whether rules that do not employ fuzzy logic (rules that employ only crisp logic) are displayed in the rule list. The rule list display shows all rules in the selected rule group when the Fuzzy and Crisp checkboxes are selected.

The Rule Editor, shown in Figure 6.11, is used to configure individual rules and optimizer functions. The Rule Editor user interface window contains several data fields and widgets that are described next.

The Rule Editor Text Field

The text field at the top of the window provides text-editing capabilities to configure expert system rules and optimizer functions. The user clicks in the text field to position the cursor appropriately and then uses the keyboard or selects from the context-sensitive entries in the Keywords, Intrinsic Functions, and Context Options lists to configure valid rule and function syntax. As text is added to the rule text editor, KnowledgeScape parses the text and updates entries in the Keywords, Intrinsic Functions, and Context Options lists. When finished, the user accesses the right mouse-button pop-up menu options to accept the rule or function (Accept Rule), accept the rule and close the grammar assistant (Accept and Close), or accept the rule as a new rule (Accept as New). The right mouse menu also allows copy, cut, paste, find, replace, and undo for editing the text of the rule.

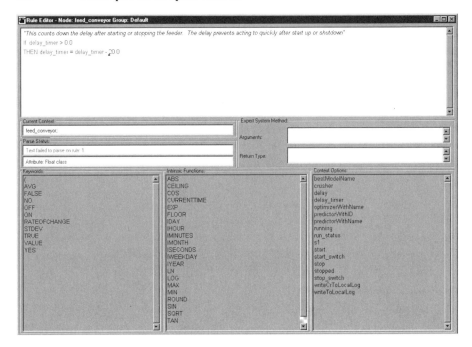

Figure 6.11   Context-sensitive Rule Editor.

The Current Context Field

The Current Context field displays the name of the KnowledgeScape object that provides the context for the rule. As the rule is created the current context will indicate the context of the last typed node or attribute. When configuring functions for an optimizer the current context field will initially display the name of the optimizer. The current context will update as the rule or function is edited. If the text "tank level" is entered into the rule text editor widget, the level attribute of the tank node will become the current context and all operations in the selection lists will be based on that context.

Parse Status

Two indicator fields display the parse status. The upper indicator displays whether the text contained in the rule text editor was successfully parsed. Parsing occurs interactively as each character is typed, and if the current text is invalid or incomplete the upper indicator will display the statement number that needs to be modified to make the rule valid. The lower indicator displays hints and information about the expected argument types for assignments and conditions, and expected input types to complete the rule or function being edited.

Methods

The Expert System Method Arguments field indicates the expected argument types for references to intrinsic and context option functions that accept arguments. For example, if "COS(" is entered into the rule text editor widget, this field will display "Number," indicating that one argument is expected that is a number.

Return Type

The Expert System Method Return Type field indicates the expected return type for references to intrinsic and context option functions. For example, if "CEILING (3.4)" is entered into the rule text editor widget, this field will display "Number," indicating that the intrinsic function returns a number.

Keywords

The Keywords list displays all keywords that are valid to complete the rule or function. As text is entered into the rule text editor widget, the keywords list is updated to display only valid entries. If the character "s" is typed, only the selections that begin with "s" are displayed. Double-clicking on an entry in the keywords list causes the selected entry to be added to the rule text editor.

Intrinsic Functions

The Intrinsic Functions list displays all valid intrinsic functions available to complete the rule or function. As text is entered into the rule text editor widget, the intrinsic functions list is updated to display only valid entries. If the character "c" is typed, only options that begin with "c" are displayed. Double-clicking on an entry in the intrinsic functions list causes the selected entry to be added to the rule text editor.

Context Options

The Context Options list displays valid options available to the current context. Entries in this list include the names of attributes, connections, subnodes, functions, etc., within the current context. As text is entered into the rule text editor widget, the context options list is updated to display only valid entries. If the character "c" is typed, only options that begin with "c" are displayed. Double-clicking on an entry in the context options list causes the selected entry to be added to the rule text editor.

Determining Node Hierarchy

The hierarchy of nodes in the process drawing determines the scope of the node in relation to other nodes. The node's scope determines which nodes and attributes can be accessed by the node's expert system rules. The scope of a node includes every subnode, any subnode of its subnodes, and nodes to which it is connected. For example, consider the car, fuel tank, and fuel pump system illustrated in Figure 6.12.

248 Techniques for Adaptive Control

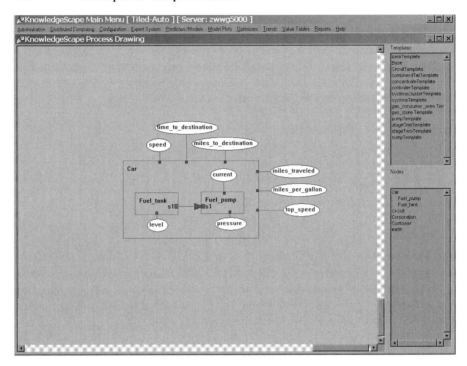

Figure 6.12 Simple representation of a car in KnowledgeScape.

The scope of the car node allows it to "see" or access the fuel tank and its level, and the fuel pump and its pressure and current. Therefore, the user can configure rules in the car node that refer to the fuel tank and fuel pump. However, the car node is not in the scope of the fuel tank node. The fuel pump is on the same hierarchy level with the fuel tank and because it is connected to the fuel tank, the fuel pump node is in the scope of the fuel tank node. The fuel pump has only the fuel tank in its scope.

The node hierarchy and its scope determine the rule syntax required to access the attributes of nodes in expert-system rules. To access attributes of the selected node, simply use the attribute's name in the rule syntax. For example, if the car node in the previous figure has attributes named "time_to_destination," "speed," "miles_to_destination," "miles_traveled," "miles_per_gallon," and "top_speed," the user would refer to those attribute directly in the rule syntax.

Examples that illustrate rules that can be configured in the car node are shown in Figure 6.13.

To access attributes of subnodes of the selected node, the user must include the name of the node before the name of the attribute in the rule syntax. For example, consider

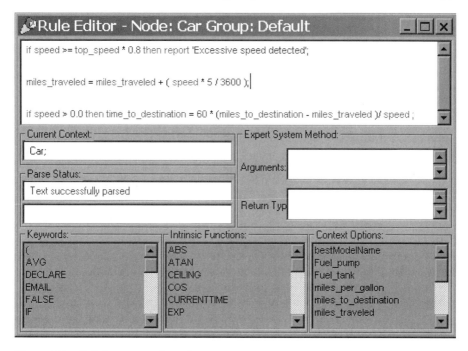

Figure 6.13 Rule browser on the node car.

the following rules that access the fuel tank level, fuel pump pressure, and fuel pump current:

> *IF fuel_tank level > 0.0 THEN miles_per_gallon = miles_traveled / (fuel_tank level *13.5)*

To access the attributes of subnodes at levels lower in the hierarchy, the user must specify the name of each node in the hierarchy, starting at the first subnode down to the target subnode followed by the attribute name as shown in Figure 6.14. The form of the rule syntax is

> *subnode_1_name ... subnode_n_name subnode_n_attribute_name*

Special Considerations in Configuring Expert Rules

The three basic KnowledgeScape crisp and fuzzy expert-system rule constructs are indicated below:

1. <Action>,
2. IF <Condition> THEN <Action>, or
3. IF <Condition> THEN <Action> ELSE <Action>

250   Techniques for Adaptive Control

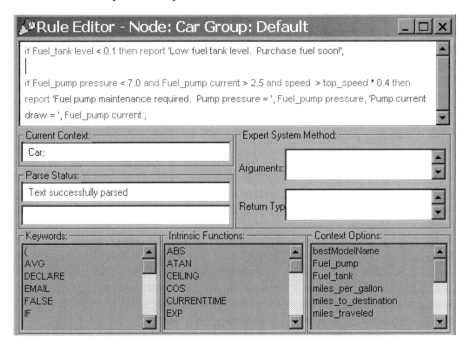

Figure 6.14   Rule editor for the car-fuel-pump example.

Any action that can be configured in an IF–THEN or IF–THEN–ELSE rule can be configured without any conditions. This is analogous to configuring a rule where the condition is always true. Some of the most common types of actions are:

- *Crisp equation actions.* Actions that calculate values and store the resulting value in an attribute of a node are equations are referred to as crisp rules. Crisp relationships typically contain an equals sign (=). For example, the following rule has the form of an equation action:
  miles_traveled = miles_traveled + (speed *5 / 3600)
- *Fuzzy equation actions.* Fuzzy actions calculate values and store the resulting value in an attribute of a node using fuzzy-logic mathematics. Fuzzy-logic equations contain the IS keyword. For example, the following rule has the form of an equation action:
  speed IS fast
- *Report actions.* Report actions generate information messages that are reported to information logs. For example, the following rule is a report action:
  REPORT *'Fuel pump maintenance required. Pump pressure =', fuel_pump pressure, 'Pump current draw =', fuel_pump current*
- *Invoke actions.* Invoke actions cause other rule groups to execute. For example, the following rule is an invoke action:

INVOKE fuel_pump high_pressure_strategy
- *Send actions.* Send actions send data to an external interface to be communicated to an external system. For example, the following rule is a send action:

  SEND fuel_pump current
- *Update actions.* Update actions cause an internal program to run, such as one that recalculates histogram values:

  UPDATE histogram_level

Boolean Considerations for Rules

Conditions configured in an IF–THEN or IF–THEN–ELSE rule generate a truth value. The truth value for crisp conditions will either be true or false. The truth value for fuzzy conditions is a value between 0.0 and 1.0.

The two types of conditions are:

1. *Crisp conditions.* When reduced to their most simple form, crisp conditions contain two values separated by a combination of mathematical operators. Valid operators include:
   - *Equal* (=). Test whether values are equal.
   - *Not equal* (~=) or (<>). Test whether two values are not equal.
   - *Greater than* (>). Test whether one value is greater than another value.
   - *Greater than or equal* (>=). Test whether one value is greater than or equal to another value.
   - *Less than* (<). Test whether one value is less than another value.
   - *Less than or equal* (<=). Test whether one value is less than or equal to another value.

   Statements that refer to values can be inserted for values on either side of the operator. For example, the following demonstrates the syntax of a valid crisp condition:

   miles_traveled
   >= miles_to_destination

   Crisp rules rely on the order in which they are inserted into the expert rule browser because they are fired in the order they are listed.

2. *Fuzzy conditions.* Fuzzy conditions contain a node attribute name, the fuzzy IS operator, and a fuzzy membership function name. For example, the following condition demonstrates the syntax of a valid fuzzy condition:

   fuel_pump current IS high

Fuzzy rules always are put at the top of the rules list in the expert-system rule browser. They have no need for a particular order because they are evaluated simultaneously and then their combined result is calculated.

Scheduling Rule Groups for Execution

Once rules have been configured in rule groups for the pertinent nodes in the process drawing, rule groups need to be scheduled for execution. Each rule group may be individually scheduled, or alternatively, rule groups may be executed when they are invoked by a rule in another rule group. Thus, one convenient method to control the execution of rules is to create an "administrative" group of rules that is scheduled to run at a regular frequency. Rules in the administrative group invoke the other rule groups for the appropriate action by use of the "INVOKE" keyword.

NEURAL NETWORKS

Neural-Network Models and Their Use in KnowledgeScape

Neural networks are a generic modeling concept based loosely upon a numerical analog of the parallel-computing model of the human brain. Research has shown neural networks to be capable of modeling any nonlinear, continuous relationship as long as there is cause and effect information in the modeled data (Haiwen et al., 1998). There are many references available detailing the theory, construction and use of neural networks, and it is not the intent of this chapter to present such information here. Rather, the intent is to present how neural networks are used within KnowledgeScape to provide learning and predictive capabilities.

Readers who are interested in the fundamentals of neural networks and the other artificial-intelligence capabilities of KnowledgeScape are referred to the list of textbooks in the Appendix to this chapter: Table of Artificial Intelligence Reference Texts.

The function of the neural networks is the same as for any generic modeling technique, that is, to provide the system or the user with a reliable representation of the cause-and-effect relationships that exist among various process parameters. With a robust model of the process, the user, or more often KnowledgeScape itself, can employ various search and optimization techniques to determine where setpoints should be set based on user-defined criteria.

Many of the screens used in creating neural networks in KnowledgeScape are shown in Figures 6.15 through 6.17. Since every object created within KnowledgeScape already has the built-in capability to train and use multiple neural-network models, the models need only be configured.

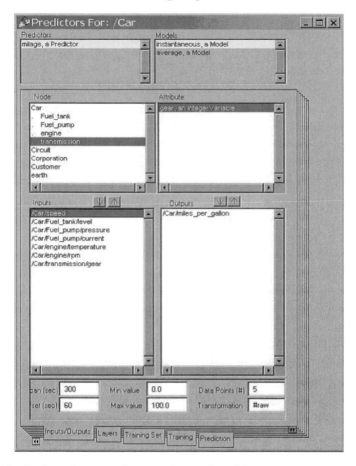

Figure 6.15 Configuration screen for a neural-network model.

Configuring Neural-Network Models

Configuration consists of selecting the attributes of various nodes that are to be used as neural-network inputs and outputs. Inputs most often consist of those process parameters that are measures; outputs typically consist of those parameters that are process-limiting, or that directly affect the economic performance of the process under the control of KnowledgeScape. Input and output data can be filtered to ensure that only values appropriate to the process are used in the training and testing of the neural network models. The topology of the network (i.e., the number and configuration of the computational nodes in the input, hidden, and output layers) can be controlled using the dialog boxes in the model-configuration screens. One such screen is depicted in Figure 6.15.

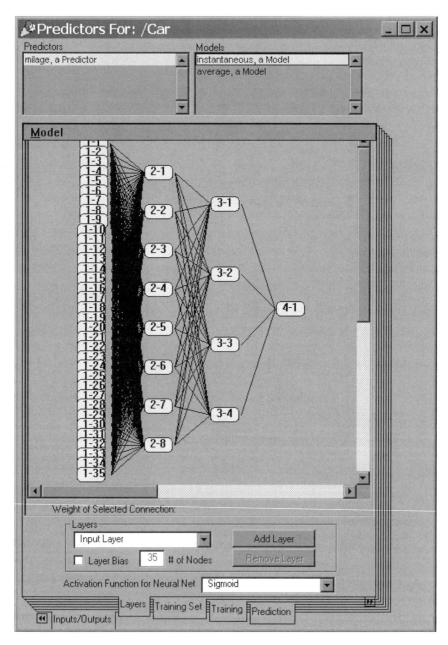

Figure 6.16 Topology of a neural network indicating the number of nodes in each layer and the sequence of connections between nodes.

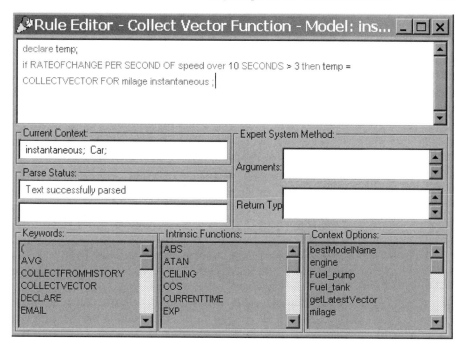

Figure 6.17 Rule editor for writing rules to filter training vectors for neural network.

How Neural Network Models are Used by KnowledgeScape

KnowledgeScape uses the concept of competitive models that vie to be deemed accurate enough to be selected by the optimization component and subsequently used to predict the near-term future state of the process. The most accurate model is selected from among the competing suite of models for use in optimization.

The modeling, prediction, and optimization functions are activated by creating a predictor, configuring one or more neural-network models for the predictor as previously described, and then starting the training and prediction functions. This is accomplished as follows.

Predictors

A predictor, with its associated models, is a collection of neural network models that have been configured to all predict the same user-selected process variables. Each model may use a different configuration for the neural network and a different combination of inputs, but each model within a predictor must have the same predicted variables, or outputs from the neural networks. If models predicted different process parameters within a single predictor object, the measure of accuracy among the competing models would be impossible to track.

Each of the individual models for a given predictor, however, may have a unique design. This concept acknowledges the reality that it is difficult to create one super model that is accurate under all processing conditions. Each model can be set up to predict continuously, and then the predictor automatically monitors their predictions and compares them with what actually occurred in the process to determine which model is actually predicting the best.

Figures 6.18 shows KnowledgeScape's predictors and neural network models in action. Predicted values are shown as the small squares while the actual process variables have been plotted as continuous traces. The first plot illustrates the adaptive nature of these models and how models improve their predictive capabilities as they gather more training information. At first the predicted values deviate significantly from the process variable, but later, after the model has been trained and updated online, the predictions match the process data much more closely.

GENETIC ALGORITHMS

The Role of Genetic Algorithms: To Use Predictors and Models for Optimization

With adaptive models that aptly demonstrate their ability to accurately predict future process states given current conditions, what additional control benefits can be achieved by using them in constrained optimization? KnowledgeScape uses genetic algorithms to interrogate the trained models as an alternate technology to determine new process setpoints.

Genetic algorithms

> are search algorithms based on the mechanics of natural selection and natural genetics. They combine survival of the fittest among string structures with a structured yet randomized information exchange to form a search algorithm with some of the innovative flair of human search. The central theme of research on genetic algorithms has been robustness, the balance between efficiency and efficacy necessary for survival in many different environments. Genetic algorithms are theoretically and empirically proven to provide robust search in complex spaces. (Goldberg, 1989)

The implementation of genetic algorithms (GAs) within KnowledgeScape is illustrated in the accompanying figures. All of the various features of a GA are available to the user through the configuration screen, including crossover and mutation probabilities, population number, and the number of generations to be used in each cycle of optimization. Suitable default values are provided by KnowledgeScape, although users may alter these values as needs arise.

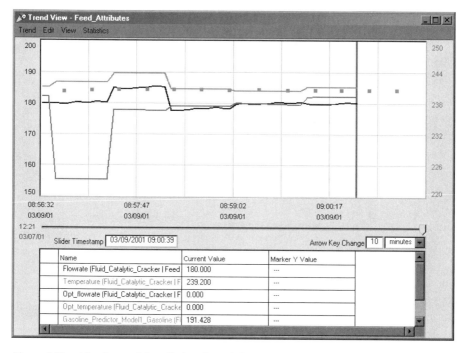

Figure 6.18 Adaptive model predictions in real time.

In the screen depicted in Figure 6.19, the uses selects which controllable process parameters the optimizer is to manipulate to achieve the desired control objective. The optimized objective is delineated within the "fitness" function. The rule-editing screen where the fitness function is defined is depicted in Figure 6.20.

Searching Model Surfaces for Best Conditions

The genetic algorithm searches the surface of the best-performing model to locate the best performance as delineated by the measure of the fitness function provided by the user. The optimizer then determines where the controlled variables need to set be to achieve the modeled optimum, and sends the setpoints matching the desired conditions to the stabilizing control system, the DCS or PLC, for implementation. This continuous, online use of competing models and optimizers is a key difference between KnowledgeScape and other expert systems.

Intelligent Optimization of Any Process

KnowledgeScape can be used to embed intelligence in any process. Process in this sense means any collection of actions that make up a system in the broadest of terms. For example, an order entry process (Figure 6.21) has customers, inventory, ship dates, price lists, etc. Business practices can be captured in expert-system rules

258 Techniques for Adaptive Control

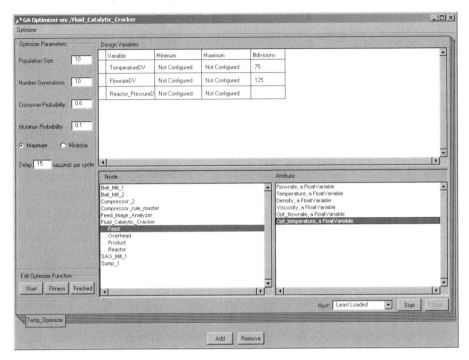

Figure 6.19 Basic configuration of a genetic-algorithm optimizer in KnowledgeScape.

describing how all of these entities are handled, how they interact, and what the desired outcomes are. KnowledgeScape can be easily used to diagram the process, specify each element of the process, and then control the process as attributes of the elements change during the process.

DOCUMENTING THE PERFORMANCE OF INTELLIGENT SYSTEMS

Documenting performance is key to the success of any control project. Knowledge-Scape has extensive reporting, trending, e-mailing and statistical-analysis capabilities built in. Rigorous techniques for documenting expert-system performance have been summarized elsewhere by Gritton (2001), but the built-in historical and statistical functions within KnowledgeScape provide an extremely useful method to graphically depict process performance.

Documenting Performance Using Histograms

Histograms are a very powerful way to keep track of performance. It is possible to write rules that govern when data is actually collected and put into a histogram. In this way it is possible to easily track performance as a function of what control strategy is being run or based on any other specific operating conditions. Figure 6.22

KnowledgeScape for the Process Industries 259

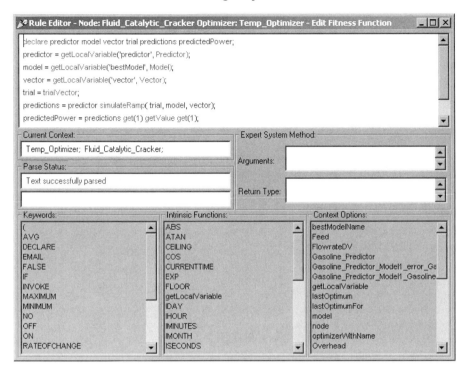

Figure 6.20 Creation of the fitness function that is either minimized or maximized by the genetic algorithm.

depicts the use of a histogram to graphically illustrate performance differences between different control techniques on one particular project.

Histograms represent the frequency (y-axis) with which a particular process variable reached a given value (x-axis). By examining the relative position along the horizontal axis one can visually compare the performance differences between competing control philosophies. When the mean and standard deviations are also known for the distributions in the histograms, rigorous statistical evaluation of the performance differences can be calculated.

Process and Attribute Trending

Trends of process data, setpoints, model predictions and optimized setpoints are commonly used to monitor the process as well as the performance of the control strategies (Figure 6.23).

KnowledgeScape also has the ability to depict model predictions in a three-dimensional view. Such depictions are invaluable in understanding complex interactions between process parameters. These graphs also assist users in understanding the

Techniques for Adaptive Control

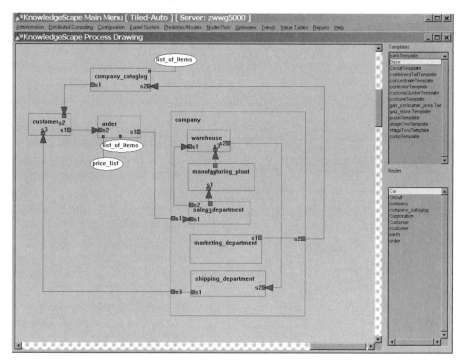

Figure 6.21 An order-entry process configured within KnowledgeScape.

effects that changes in the control variables will have on performance. Figure 6.24 depicts a three-dimensional representation of a multidimensional neural-network model surface.

Communicating Performance Measures

Communicating performance measures is as important as calculating such metrics. KnowledgeScape has an integrated report writer that can be used to configure standard or special reports. Once configured and saved, these reports may be generated at any user-prescribed interval and mailed electronically using standard e-mail protocols.

Figure 6.25 depicts the user interface used to format reports. Any node or attribute, including histograms, charts, and expert rules, may be included in such reports. In addition, data can be filtered, formatted, and depicted in a multitude of formats using the report writer.

Once configured and saved, reports can be printed or scheduled for e-mail delivery using the report-scheduling interface depicted in Figure 6.26. User-selected reports are e-mailed at the specified times to the addresses entered in the text field in the lower portion of the screen.

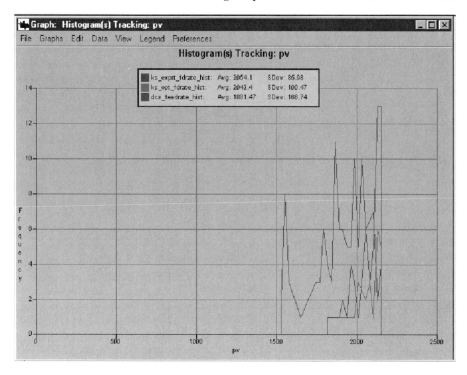

Figure 6.22   Histogram configuration and histogram.

PUTTING IT ALL TOGETHER: COMBINING INTELLIGENT TECHNOLOGIES FOR PROCESS CONTROL

The intelligent integration of the components within KnowledgeScape has been done for the express purpose of providing the user with an easy-to-use, seamlessly integrated control and/or modeling package. The sequence of steps needed to successfully configure and implement a KnowledgeScape system are summarized as follows:

1. Configuring an Application in KnowledgeScape

The involves configuring the various nodes on the process drawing and attaching the proper attributes. Remember that nodes can represent any physical object or any defined process. Attributes are the measurable descriptors that we use to describe or control the behavior or performance of the objects represented by the nodes.

Data servers should also be configured for any nodes that communicate to external systems such as the plant control systems and instruments.

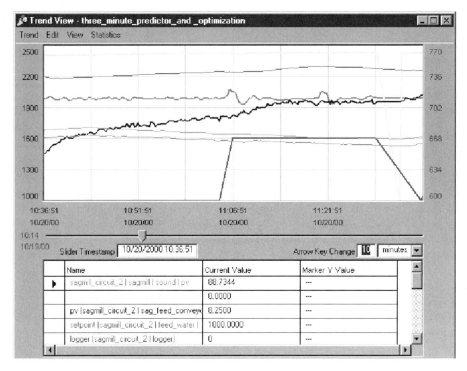

Figure 6.23 Trending process attributes to monitor performance.

2. Writing Crisp and Expert Control Rules

Once nodes have been configured with appropriate tags or attributes and data servers configured, the user then can write the expert-system rules. These rules codify best known practices for the application.

To enable the user to write rules in fuzzy syntax, the appropriate fuzzy sets must be configured. A right mouse click on the attribute in question will allow the user the option to configure the fuzzy sets for any attribute on the process drawing. Once these sets are configured using the graphical interface, the rule editor will allow the user to write rules in fuzzy syntax.

3. Creating Adaptive, Online Neural Models of the Process

Having configured the process modes and their attributes, users may configure neural-network models. Note that a model can be configured, trained, and tested and predictors configured and started even if the expert-system rules are not written or used. The only dependency is on the existence of the appropriate nodes and attributes in the process drawing. Use of all of the intelligent components in KnowledgeScape is optional.

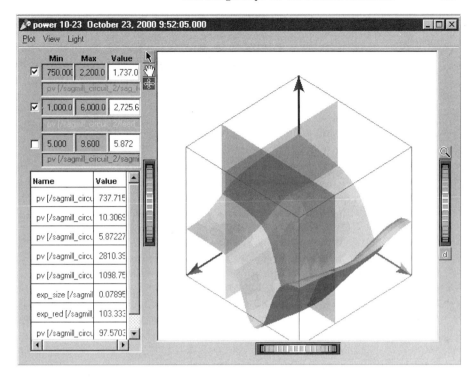

Figure 6.24   A three-dimensional view of the surface of a multidimensional neural-network model.

Neural networks are configured to model the process and predict the process performance by right-clicking inside the node containing the attributes to be modeled or predicted. A graphical interface with lists of available process attributes is used to configure the models. Separate tabs in the model-configuration window allow the user to control the amount of time to be spanned by the models, the method and frequency of data collection, model training, and model predictions. The user may also select which networked computers run the models; alternatively, Knowledge-Scape can assign the models to the "least loaded" computer.

4. Creating Genetic Algorithm Optimizers for the Process

With rules, models, and predictors in place, the user may configure optimizers to determine the best set of process conditions based on the user's own definition of "best." The optimizer is configured with default of user-changeable parameters for the genetic algorithms, and the user also provides the definition of the condition to be maximized or minimized by the optimizer. This is defined within the fitness function. Once the optimizer is running, KnowledgeScape will search the appropriate regions for the pertinent controllable process parameters and automatically optimize process performance.

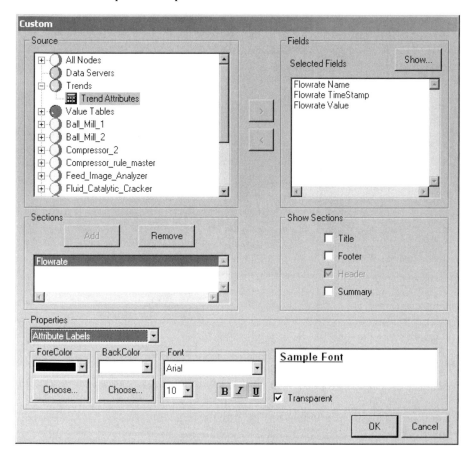

Figure 6.25 Configuring reports using the report writer.

5. Using the Built-In Performance-Documentation Features

Once the system is configured and online, the user can take advantage of the built-in documentation and performance-measuring tools. Specifically, the report writer, the report scheduler and mailer, the histograms, and the intrinsic statistical functions within KnowledgeScape are all there to assist the user in quantifying the benefits of the control system.

RESULTS: USING INTELLIGENT CONTROL IN THE MINERAL-PROCESSING INDUSTRY

KnowledgeScape has demonstrated its inherent value when applied in an online process-control application. Experience has demonstrated that the more consistently KnowledgeScape is used online, the greater the overall benefits to the plant or process. Typical industry-wide results for mineral-processing applications are:

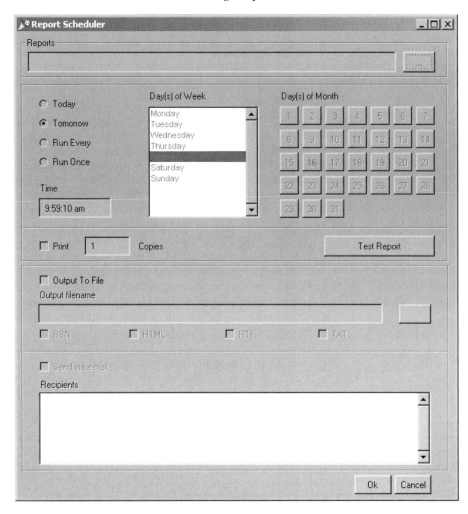

Figure 6.26 User interface for scheduling reports to be e-mailed to specified addresses.

- 4–8% improvement in grinding throughput at the same or similar grind size
- 1–3% increase in flotation recovery
- 2–8% improvement in concentrate grade
- 10–40% decrease in reagent consumption

Note that these improvements are not exclusive. For example, it is common to see improvements in both concentrate grade and recovery. Grade must be traded for recovery if a process is operating at a thermodynamic limit, but plants typically operate at something less than the thermodynamic optimum. Hence, under plant-wide expert control, concurrent improvements in grinding throughput, recovery,

Table 6.1 Examples of Project Benefits

Mineral	Expert Control Scope	Results
Copper	12 lines, ball milling	12% increase in production
Iron ore	13 lines, ball milling	11% increase in production
Copper–gold	5 lines, ball milling	12% increase in production
Gold ore	SAG-mill, ball-mill circuit	11% increase in production
Copper	SAG-mill, ball-mill circuits, and flotation	8% increase in production
Gold–lead–zinc	SAG-mill, ball-mill circuits, and flotation	5% increase in production
Copper	SAG-mill, ball-mill and crushing circuits	5% increase in production
Copper	10 lines, ball milling	18% increase in production
Gold	SAG-mill, ball-mill circuits	9% increase in production

grade, and reagent consumption can be and are typically realized. Typical results from a number of projects are summarized in Table 6.1.

KnowledgeScape makes literally thousands of setpoint changes in a day as it relentlessly applies the best possible control logic to the ever-changing conditions in the mill. It does what human operators cannot—that is, constantly analyze performance and apply the best known control logic to maximize the performance of the mill. It does this without ever getting tired, confused, or distracted.

Specifically, KnowledgeScape systems used to control grinding and flotation plants have significantly increased operating revenues. We have learned over time that the control technologies integrated into KnowledgeScape have incremental effects on the improvements achieved. For example, in grinding control, improvements shown in Table 6.2 are typical.

CONCLUSION

KnowledgeScape is the first process-control system that has integrated the powerful capabilities of real-time expert control, online adaptive and competitive neural network models, and genetic algorithm optimization into one software system designed to monitor and control any industrial process. Careful reflection on the descriptions of functions of KnowledgeScape suggests that its design can be used in a much broader sense. Specifically, KnowledgeScape can be used to embed intelligence in any process.

Table 6.2 Improvements Associated with Different Artificial Intelligence Technologies

Functionality	Incremental Improvement
Crisp heuristic rules	4%
Fuzzy heuristic rules	2%
Model-based optimization	1–2%

Artificially intelligent tools integrated into KnowledgeScape easily allow forces on economic optimization of any process. The following are some of the tools included.

Crisp and Fuzzy Expert System

These components allow the user to write rules in an English-like syntax that cause the expert control system to mimic the actions that would be taken by a very expert operator. These components provide the first layer of intelligence for the system and track emergency conditions, control the execution of the control strategies, track system performance, and determine when to use the recommendations made by models and optimizers.

Neural Network Models

The neural networks allow the expert system to learn about process dynamics that may or may not have been anticipated when the expert rules were written. They provide models of the process for use in process optimization, in visualizing the interactions of process parameters, and in visualizing the effects of process parameters on process performance.

Optimizers

The role of the optimizers is to identify which process models most accurately predict process performance under a given set of operating conditions. The optimizers then search the multidimensional surface of the best process model to determine what combination of process setpoints will yield incrementally higher performance.

In addition, the optimizers allow the user to determine what process-control objective is most important. This objective is written in a rule editor, much the same as configuring expert-system rules. The difference is that this rule, or set of rules, defines the most desirable state of operations.

Report Writer

The report writer and the associated report scheduler track the performance of the system as requested by the user. The module will e-mail preconfigured reports using simple mail transport protocol (SMTP). The report writer is also used to document the expert system and provides the user a way to print out and evaluate any or all rules, models, process variables, or any other attributes of the expert system.

REFERENCES

Goldberg, David E. (1989). *Genetic Algorithms in Search, Optimization and Machine Learning.* Addison-Wesley, Reading, MA.

Gritton, K. S. (2001). "Methods to Document the Benefits of Advanced Control Systems." SME Preprint 01-20, The Metallurgical Society, Littleton, CO.

Haiwen, Y., R. Nicolai, L. Reh (1998). "A Bayesian-Gaussian Neural Network and Its Applications in Process Engineering—Modelling, Optimization, and Control." *SPEEDUP J.* **12** (1).

Hales, L. B., R. A. Ynchansti, D. G. Foot, Jr. (2000). "Adaptive Object-Oriented Optimization Software System." U.S. Patent No. 6,112,126 (August 29).

APPENDIX: TABLE OF ARTIFICIAL INTELLIGENCE REFERENCE TEXTS

Author(s)	Title	ISBN	Subject*
Martin Anthony, Peter L. Bartlett	*Neural Network Learning: Theoretical Foundations*	052157353X	NN
Wolfgang Banzhaf, Colin Reeves	*Foundations of Genetic Algorithms*	1558605592	GA
Dimitri P. Bersekas, John N. Tsitsiklis	*Neuro-Dynamic Programming* (Optimization and Neural Computation Series, 3)	1886529108	NN
Lance Chambers	*Practical Handbook of Genetic Algorithms*	0849325196	GA
Lance Chambers	*The Practical Handbook of Genetic Algorithms: Applications*, 2nd ed.	1584882409	GA
Earl Cox	*The Fuzzy Systems Handbook: A Practitioner's Guide to Building, Using and Maintaining Fuzzy Systems*	0121944557	FL
David E. Goldberg	*Genetic Algorithms in Search, Optimization and Machine Learning*	0201157675	GA
Martin T. Hagan, Howard B. Demuth, Mark H. Beale	*Neural Network Design*	0534943322	NN
Randy L. Haupt, Sue Ellen Haupt	*Practical Genetic Algorithms*	0471188735	GA
Simon S. Haykin	*Neural Networks: A Comprehensive Foundation*	0132733501	NN
L. C. Jain, N. M. Martin	*Fusion of Neural Networks, Fuzzy Sets and Genetic Algorithms: Industrial Applications* (Int'l Series on Comp. Int.)	0849398045	NN, GA, FL
Michael J. Kearns, Umesh V. Vazirani	*An Introduction to Computation Learning Theory*	0262111934	NN
George J. Klir, Bo Yuan	*Fuzzy Sets and Fuzzy Logic: Theory and Applications*	0131011718	FL
Anders Krogh, et al.	*Introduction to the Theory of Neural Computation*	none	NN

(*Continued*)

TABLE OF ARTIFICIAL INTELLIGENCE REFERENCE TEXTS (*continued*)

Author(s)	Title	ISBN	Subject*
Melanie Mitchell	*An Introduction to Genetic Algorithms* (Complex Adaptive Systems Series)	0262631857	GA
Hung T. Nguyen, Elbert A. Walker	*A First Course in Fuzzy Logic*	0849316596	FL
Witold Pedrycz, Fernanco Gomide	*An Introduction to Fuzzy Sets: Analysis and Design (Complex Adaptive Systems)*	0262161710	FL
Jose C. Principe, Neil R. Euliano, W. Curt Lefebre	*Neural and Adaptive Systems: Fundamentals through Simulations*	0471351679	NN
Russell D. Reed, Robert J. Marks II	*Neural Smithing: Supervised Learning in Feedforward Artificial Neural Networks*	0262181908	NN
S. Tambe, B. D. Kulkarni, P. B. Deshpande	*Elements of Artificial Neural Networks*	0965163903	NN

*NN, neural networks; GA, genetic algorithms; FL, fuzzy logic.
Reprinted from Kenneth S. Gritton, Methods to Document the Benefits of Advanced Control Systems, Society of Mining and Metallurigcal Engineers. Preprint 01–20, February 2001.

AUTHOR INDEX

Anderson, B. D. O., 155, 201
Åström, K. J., 14, 21, 53, 54, 56, 97, 109, 142, 201, 203, 206, 207, 208, 232

Bakshi, B., 206, 232
Bartee, J. F., 97
Box, G. E. P., 213, 232
Bristol, E. H., 24, 49, 53, 168

Cheng, G. S., 145, 179, 193, 201, 202
Chia, T. L., 203, 206, 217, 232
Clarke, D. W., 56, 97, 105, 143, 202
Cohen, G. H., 206, 232
Coon, G. A., 206, 232

de Dona, J. A., 112, 143
Desforges, M. J., 97
Desoer, C. A., 155, 201
Dumont, G. A., 99, 102, 105, 109, 116, 143

Foot, D. G., Jr., 268
Fu, Y., 102, 143, 202

Gawthrop, P., 56, 97
Geladi, P., 57, 97
Georgakis, C., 97
Goldberg, D. E., 256, 267, 268
Goodwin, G. C., 109, 143, 202, 206, 232
Gough, W. A., 99
Goulding, P. R., 97
Gritton, K. S., 233, 258, 268, 269
Gustavsson, I., 206, 232

Hagglund, T., 53, 201, 206, 232
Haiwen, Y., 252, 268
Hales, L. B., 233, 268
Hang, C. C., 168, 201
Hansen, P. D., 23, 24, 33, 35, 53, 54

Hastings, J. R., 55, 97
Haykin, S., 56, 97, 268
He, M., 201
Head, J. W., 101, 143
Hesketh, T., 56, 97
Ho, W. K., 201
Horch, A., 143
Huzmezan, M., 99

Isaksson, A., 116, 143

Jenkins, G. M., 213, 232

Kalman, R. E., 3, 21, 112
Kammer, L., 116, 143
Kuo, B., 58, 97, 202
Kurth, T., 97

Lee, D. T., 232
Lee, T. H., 201
Lee, Y. W., 101, 143
Lefkowitz, I., 203
Lennox, B., 97
Li, D. L., 201
Ljung, L., 64, 97, 206, 232

Marin, O., 197, 201
McFarlane, R. C., 76, 77, 97
Middleton, R. H., 143
Mohtadi, C., 105, 143
Morari, M., 202, 232
Morris, C., 191, 201
Murrill, P., 201

Nichols, N. B., 21, 232

Panagopoulos, H., 25, 53
Pao, Y. H., 232

Pollard, J., 206, 232

Qin, S. J., 56, 61, 97

Reineman, R. C., 97
Rivera, D. E., 208, 217, 232

Sage, M. W., 55, 97
Salgado, M. E., 104, 143
Sandoz, D. J., 55, 56, 59, 97
Seiver, D., 197, 201
Seron, M. M., 143
Shinskey, F. G., 29, 32, 53, 201, 202
Sin, K. S., 109, 143, 202, 206, 232
Skogestad, S., 232
Sobacic, D. J., 207, 232

Soderstrom, T., 206, 232
Swanick, B. H., 55, 97

VanDoren, V. J., 1, 189, 201
Vidyasagar, M., 155, 201

Wittenmark, B., 14, 21, 54, 56, 97, 109, 142, 201, 206, 207, 208, 232
Wong, O., 59, 97
Woolley, I. S., 97

Ynchansti, R. A., 268

Zervos, C. C., 101, 105, 109, 143
Ziegler, J. G., 5, 21, 206, 232

SUBJECT INDEX

Actuators. *See* Control loop fundamentals
Adaptation, 5, 8–9, 12, 16, 23, 51, 66, 72, 117, 189, 191, 204–205, 235
Adaptive control, 2–20, 23, 48, 56, 58, 70, 88, 91, 96, 99, 100–103, 105, 108–109, 114–116, 127, 129, 134–135, 137, 140–142, 145–153, 158–159, 163–164, 170–173, 186, 188–191, 196–197, 201, 204–209, 211, 213–214, 216–220, 223, 225, 230–231
Adaptive Resources, 3
Air separation unit, 197–198
Algorithms, 2, 13–14, 50, 65–66, 101–102, 104–106, 109, 111–112, 114, 116, 140, 142, 148–149, 152–153, 175, 184, 186, 188–189, 206, 208, 210, 213, 216–217, 231, 233, 235, 240, 256–257, 259, 266
ARMA/ARMAX. *See* Models and modeling
Artificial intelligence, 235, 239
 expert systems, 8–10, 15–17, 20, 209, 217, 219, 230, 234–236, 240–245, 247, 251, 257, 260, 267
 fuzzy logic, 202, 245, 268–269
 genetic algorithms, 233, 235, 240, 256–257, 259, 263, 266, 269
 rule-based control, 16–18
 rule sets, 9, 17
ARX. *See* Models and modeling
Automation, 7, 64, 72, 75, 108, 114, 132–133, 137, 158, 170, 180, 182, 184, 190, 194–195, 203–204, 206, 229
Auto-tuning. *See* Tuning

Bachelor Controls, 3
BciAutopilot, 3
Black box control, 7, 101–102, 146, 153, 188, 230
Bode plots, 27, 32–34, 39, 41–43, 46–47

BrainWave, 3, 6–8, 11–12, 14, 16–18, 20–21, 99–101, 103, 105–107, 109, 111, 113, 115–117, 119–125, 127, 129, 130–133, 135, 137, 139–141, 143

Cascade control, 173–177, 182
Catalytic crackers, 57, 76–77, 83
Chemical processes, 1, 60, 62, 75, 101, 117, 126, 130, 182–184, 195–196, 215
Closed loop. *See* Control loop fundamentals
Coking furnace, 192–194
Commissioning, 3, 71, 102, 118–119, 178, 198
Composition control, 23–24, 48, 62, 75, 122, 132, 193
Connoisseur, 3, 6–8, 11–14, 17–18, 20–21, 55–57, 63, 66, 68, 70–75, 78, 81, 96
Constraints. *See* Control loop fundamentals
Continuous processes, 1, 49, 72, 103, 117, 122, 150, 152, 155, 191, 194, 221, 225, 234, 252, 256–257
Control effort/control action. *See* Control loop fundamentals
Control engineers/engineering, 1, 2, 9, 55, 57, 59, 61, 64, 67, 71, 102, 195, 209
Control horizon. *See* Control loop fundamentals
Control law/control equation. *See* Control loop fundamentals
Control loop fundamentals
 actuators, 2, 12, 15, 17, 139–140, 180, 182
 closed loop, 1, 5, 14–15, 20, 28, 32, 35, 37, 39, 43–46, 48, 52–53, 69–70, 72, 87, 96, 100, 115–116, 118, 120–125, 132, 134, 137, 148–149, 153–154, 168, 174, 190, 205 206, 208–209, 216, 218–219, 224
 constraints, 25, 28, 31, 44, 52, 62, 91, 93, 96, 102, 106, 112–113, 118, 125, 127, 129, 132, 142, 185, 225, 231, 243, 256

273

274 Subject Index

Control loop fundamentals (*Cont.*)
 control effort/control action, 7–9, 14–16, 19, 70, 109, 116, 120, 123, 133, 139–140, 157, 167, 170, 175, 183, 186, 190, 212
 control horizon, 106, 113
 control law/control equation, 2–3, 7–9, 11, 13–15, 109–110, 113, 116–117
 control loops/control systems, 3–5, 8, 11, 17, 23–25, 27–28, 31, 33–34, 36–38, 43, 45–46, 48, 51–53, 57, 63–64, 67–71, 75, 83, 90, 96–97, 102, 111, 114–115, 129, 131–132, 134–135, 137, 139–140, 146, 148–150, 153–155, 158–161, 164, 167–170, 173–178, 180, 182–184, 186, 190–191, 193, 195–196, 203–205, 209, 216–220, 222–223, 225–227, 229–231, 234–235, 240, 257, 261, 264, 267
 error, 1–3, 5, 8, 14–15, 17, 26–27, 30–31, 35, 39, 46, 52–53, 68, 72, 105, 109, 111, 120, 147–148, 150, 162–163, 165, 189–190, 226–227
 feedback, 1, 5, 23–27, 33, 35, 37, 41, 44–48, 50–52, 57, 67, 70–71, 96, 106, 109, 112, 119, 120–121, 123–124, 149–150, 154–155, 158–159, 164, 170–171, 177–178, 184, 189–190, 195, 209, 216, 231
 feedforward, 20, 23–27, 47–48, 50–52, 61, 71, 83, 105, 109–111, 113, 118–119, 121–123, 140, 147, 157, 177–179, 183–184, 192, 195
 manipulated variable/controller output, 2–3, 7, 9, 12, 14–15, 17, 23–26, 30–33, 35–36, 39, 43–52, 56, 58, 61, 63, 71, 78–80, 83, 86, 88–89, 96, 101, 109, 114, 120, 147, 150, 157–158, 160, 167–168, 170, 174, 177, 183, 193, 209, 213, 216, 219, 226, 240
 measurements, 10, 12, 23, 25–26, 28–30, 32, 34–39, 44–47, 52–53, 59, 64–65, 109, 119, 122, 189–190, 215
 noise, 10–12, 25, 28, 31, 34–35, 49, 60, 62, 64–65, 70, 76–77, 80, 100, 109, 115–116, 147, 150, 208–211, 215–216, 218, 221, 225
 objectives, 7, 9, 20, 27, 63, 83, 106, 129, 160–161, 192, 195, 216, 235, 241
 open loop, 27–28, 34–37, 43–46, 51–53, 62, 67–70, 72, 96, 115, 117–118, 124, 132, 137, 149, 153–156, 168–169, 190
 process input, 2, 12, 19, 112, 148, 153, 157, 167, 180–181, 188, 190, 196, 206, 209, 216
 process output, 2, 9–12, 19, 111, 148, 157, 167–168, 181, 215–216
 reference trajectory/value, 2, 112
 sampling, 14, 44, 59–62, 68, 70, 81, 83, 100, 102, 106, 109, 112, 115, 118, 120, 142, 157, 166, 208–209, 212–213
 sensors, 1, 26, 191, 195, 226, 230
 setpoint, 2, 5–7, 10–12, 14, 17, 25–27, 29–32, 35–36, 38–39, 41, 43–44, 46–47, 51–53, 75, 77–78, 83, 90, 92–94, 96, 105, 112, 114, 116–127, 129–130, 132–134, 137, 139–141, 147, 150, 158, 160, 174–177, 185, 189, 196–197, 203, 205, 207, 212–214, 216, 218–222, 224–227, 229, 231, 266
Control theory, 4, 8, 15, 149, 180, 187, 206, 209
ControlSoft, 3, 206, 217
Cooling, 15–16, 117, 126–130, 138–139, 184
Coupling. *See* Process behavior
Curve-fitting, 10, 13, 19
CyboCon, 3, 6–8, 14–15, 17–21, 147, 149, 151, 153, 155–157, 159, 161, 163, 165, 167, 169, 171–173, 175–179, 181, 183, 185, 187, 189, 191–193, 195–197, 199, 201–202
CyboSoft, 3, 201–202

Damping. *See* Process behavior
Detuning. *See* Tuning
Diagnostics, 219, 229–231
Distillation, 67, 75
Disturbance rejection, 114, 135, 231

EXACT, 3, 7–8, 10–11, 17, 20–21
Expert systems. *See* Artificial intelligence

Feed rate/feed rate, 78, 127–129
Filtering, 12, 25, 28, 30, 32, 36–39, 43, 49, 62, 64–66, 70–71, 76, 87–88, 103, 112, 115–117, 125, 151–152, 218, 253, 255, 260
Finite impulse response. *See* Models and modeling
Finite step response. *See* Models and modeling
First principles. *See* Models and modeling
Flow control, 47, 78, 117, 130, 137, 175
Flow rate, 1, 119, 122, 193
Food, 9, 126, 191–192

Subject Index

Foxboro Company, The, 3, 23, 25–27, 29, 31, 33, 35, 37, 39, 41, 43, 45, 47, 49, 51, 53, 55, 206
Fuzzy logic. *See* Artificial intelligence

Gain scheduling, 5–6, 23–24, 26, 48, 51–53, 186, 205
Genetic algorithms. *See* Artificial intelligence
Glass, 137–142

Heat/heating, 3, 67, 76, 117, 126, 130–132, 138, 170, 184, 204–205

Initial conditions, 154–155
Internal model control, 111, 217, 231
INTUNE, 3, 8–9, 15–17, 20–21, 203, 206, 208, 217–227, 229–231
Inverse response. *See* Process behavior

KnowledgeScape, 3, 8–9, 15–17, 20–21, 233–237, 239–243, 245–249, 251–253, 255–267, 269

Laguerre functions, 6, 13, 99–105, 107–110, 115, 117, 142–143
Laplace transforms, 49, 58, 101, 146, 178, 217
Lead/lag control, 111
Least squares. *See* Models and modeling
Level control, 117, 154, 156, 185, 190, 198, 225
Linear process/model/control, 19–20, 25, 31, 34, 44, 48, 53, 57, 72–74, 80, 100–106, 109, 115, 154, 170, 176, 180–181, 190, 206
Linear quadratic Gaussian control, 31, 33

Maintenance, 20, 71, 96–97, 118, 178, 192, 198, 225–226, 243, 250
Manipulated variable/controller output. *See* Control loop fundamentals
Manual tuning. *See* Tuning
Margins
  frequency, 31, 33, 46
  gain, 33
  phase, 20, 33, 35, 68
Mechanical process, 1, 136, 139
Mining, 266
Model-free control, 8, 15, 19, 145–165, 167, 169–199, 201–202
Model predictive control, 55, 57, 69–72, 75, 77–78, 83, 88, 90, 96–97, 99, 105–106, 110, 112, 118, 208

Models and modeling, 10–12, 14, 17–19, 30, 55–57, 63, 66–75, 83–87, 90–97, 99–103, 112, 115–116, 119, 142, 191, 208, 211, 215, 218, 234, 238, 252, 255, 261
  ARMA/ARMAX, 13, 77, 102–103, 215
  ARX, 56, 62–65, 71, 76, 78, 83, 88–95
  differential equations, 9, 10, 49, 59, 61, 146, 154–155
  errors, 30, 69, 215
  finite impulse response, 59, 61, 63–65, 68, 71, 75, 78–81, 83–86, 88, 89–90
  finite step response, 59, 61, 64, 68, 81
  first order plus deadtime, 115, 120–121, 215, 217
  first principles, 10, 100, 188
  identification, 7, 9, 10, 12, 14, 18–19, 43, 45–46, 51–52, 55, 57, 60, 61–71, 73–76, 78–81, 84, 87–92, 96, 100–102, 105, 108–112, 115–117, 119–123, 145–149, 191, 206, 208–219, 222–223, 230–232, 240, 267
  least squares, 13, 56–57, 65–67, 70, 72, 75–76, 83, 88, 91, 97, 100–101, 104, 106, 109, 206, 208, 210–211
  maximum likelihood, 208, 211
  parameters/coefficients, 8–10, 49–50, 57, 63, 66–69, 71–72, 78, 81, 83, 86, 92, 94, 109, 115, 206, 211, 215–216
  persistent excitation, 10, 16, 68, 146–147, 149, 216–217
  prediction horizon, 83, 105–106, 120, 124
  process/model mismatch, 5, 25, 33, 43, 46, 172
  radial basis functions, 6, 56–57, 73–75
  state-space, 59–60, 62, 68, 101, 103, 170, 231
  transfer functions, 29, 35–36, 45, 47, 49, 58–61, 63–64, 68, 70, 82, 88, 102–104, 117–123, 146, 178
Motors, 1, 236
Multivariable process/model/control, 20, 23–24, 55–59, 62–63, 67, 75, 88, 96, 147, 154, 158–159, 163–170, 173, 190–193

Neural networks, 6, 16, 57, 150–152, 162, 189, 233–235, 240, 252–256, 266–267, 269
Nonlinear process/model/control, 5, 10, 19, 20, 23, 56–57, 60, 62, 68, 72–74, 79, 96, 101, 109, 114–115, 139, 147, 154–155, 157, 177, 180–182, 189–191, 204–205, 208, 213–214, 252
Nyquist, 28, 33–34, 39, 41–43, 52

Subject Index

Open loop. *See* Control loop fundamentals
Optimization, 44, 67, 101, 104–106, 109, 112, 142, 175, 211, 233–235, 238, 240, 252, 255–256, 266–267
Ore grinding, 242, 266
Oscillations. *See* Process behavior

Paper mill, 134, 136–138
Pattern recognition, 18, 206–207, 209, 217–219
Performance, process/model/control, 2–3, 5–6, 9, 11, 17, 25–26, 30–31, 33, 43–45, 51, 56–57, 69, 72, 83–84, 97, 100, 105–106, 110, 112, 114, 119–121, 123–125, 127–128, 133–135, 137–138, 140–142, 146, 148, 152, 157, 167, 170–174, 177, 181, 183–184, 187, 196, 198, 204–206, 208–209, 225–226, 230–231, 234–235, 238, 240, 253, 257–263, 266–267
Persistent excitation. *See* Models and modeling
Petrochemicals, 9, 56–57, 70, 76, 78, 193–194
pH control, 20, 122, 182, 195–197
PID control
    proportional-integral, 35, 38, 43, 174, 206
    proportional-integral-derivative, 4–6, 8, 11, 15, 23, 26–30, 32–33, 35–36, 39, 43–44, 46, 51–54, 56, 68, 99–100, 111, 117, 125–127, 129–135, 139–142, 147, 154, 170–171, 174–175, 187–189, 191–192, 196, 198–201, 203, 206, 208, 211, 215, 217–222, 224–225, 229, 231–232
    proportional-only, 2–3
Power, 1, 49–50, 52, 61–62, 134, 142, 243
Prediction horizon. *See* Models and modeling
Pressure control, 25–26, 48, 76, 126–127, 130, 134, 176–177, 181–182, 191, 193, 197, 248–251
Pretuning. *See* Tuning
Process behavior, 5, 14, 23, 63–64, 170, 190, 216–221, 224, 226–227, 229
    controlled variable/process variable, 1–2, 5, 7, 12, 17–19, 24–26, 30–31, 35, 47–49, 51, 53, 58, 60, 67, 76, 80, 86, 92, 101, 111–112, 116, 147, 150, 158–160, 164, 167–168, 171, 175–176, 178–179, 181, 185, 189, 198, 207, 209, 214, 218–222, 224, 226–227, 231, 235, 255–257, 259, 267
    coupling, 24, 26, 48, 160, 163, 166–170, 183, 193, 225

    damping, 25, 35–37, 45, 58, 64, 83, 88, 120, 209, 225
    deadtime/time delay, 17, 19, 23, 25–27, 29, 31, 35–36, 44, 53, 58, 61, 64, 68–69, 88, 102–103, 107, 110–112, 114–120, 122, 124–126, 129, 131, 147, 166, 168–173, 179, 191, 193, 195, 206, 209, 213–215, 217
    disturbances/upsets/loads, 5, 7, 10–13, 17, 23–27, 31–32, 35–36, 38–39, 41, 43–49, 51–53, 67, 72, 77, 83, 104, 109, 110–114, 119, 134–135, 150, 154, 158, 175, 177–179, 183, 195–198, 203, 209, 213, 219, 222, 226, 229, 231
    dynamics, 23, 57–63, 67, 70, 79, 82, 88, 102–103, 105, 108, 110–111, 116–117, 119–120, 123, 129, 135, 139, 146, 150, 154, 174, 183, 189, 191, 206, 208–211, 213, 217–218, 220–222, 231, 267
    gain, 2–3, 5–6, 15, 20, 23–24, 26–28, 30–37, 39, 45–46, 48, 50–53, 58, 61, 64, 68, 81–82, 108, 117, 139, 148, 152, 154, 157–158, 162–163, 165, 167–170, 173–174, 178–179, 181–182, 186, 188, 196, 204–205, 210, 215, 217
    impulse response, 13, 68
    input/output, 10–14, 19–20, 28, 43, 49, 152, 154–155, 188, 190, 207
    inverse response, 19, 58
    nonminimum phase, 19, 27, 83, 88, 108
    nonself-regulating, 35, 116–117, 124–125, 154, 156, 190, 225
    oscillations, 10, 15, 45–46, 52–53, 101, 127, 132, 182, 220–222, 224–226, 229
    overshoot/undershoot, 17, 25, 30–31, 43–44, 46, 157–158, 167, 170, 209, 224, 231
    settling time, 27, 139, 141, 209
    steady-state, 12, 26, 30, 52, 61–62, 67–68, 79, 81, 88–89, 110–111, 117–118, 138, 147, 168, 203, 221
    step response, 38–39, 89, 102
    time constants/lag time, 26–31, 35–36, 38–39, 46, 52–53, 58, 61, 63–64, 68–69, 82–83, 88, 99, 102, 114–118, 124, 126, 129, 131–132, 135, 139, 157–158, 166–168, 173, 179, 188, 190, 206, 215, 217
    time-invariant, 101, 170
    time-varying, 5, 19, 154, 170, 190, 204–205, 211, 231
    transients, 26, 50, 115, 118, 123–125, 203, 209, 218–219, 221, 224

Subject Index

Process control, 1, 2, 4, 13, 16–17, 67, 99, 101, 108, 114, 126, 146–147, 170, 174–175, 207, 219, 225, 230–231, 242
Pseudo random binary sequence, 11, 56, 63, 67–68, 70–72, 96, 214, 216

Quadratic programming, 83, 106
QuickStudy, 3

Radial basis functions. *See* Models and modeling
Reactors, 117, 126–133, 183–184
Reference trajectory/value. *See* Control loop fundamentals
Regulatory control, 25, 35, 64, 99, 117, 134, 150, 160, 197
Reporting, 219, 229–230, 232, 250, 258, 260, 264, 267
Robustness, 7, 26, 31, 33–34, 44–45, 52–53, 65–66, 75, 99, 108, 123, 142, 147, 154, 157, 172, 184–185, 191, 198, 223, 230, 234–235, 252, 256
Rule-based control. *See* Artificial intelligence
Rule sets. *See* Artificial intelligence

Safety, 6, 11–12, 24, 72, 127, 198, 217, 225
Scheduling, 23–25, 53, 186, 265
Self-tuning. *See* Tuning
Sensitivity, 2–3, 20, 35, 49, 103, 108, 129, 138, 152–154, 193, 204–205, 215
Sensors. *See* Control loop fundamentals
Setpoint tracking, 27, 30, 212
Simulation, 46, 51, 57, 77, 100, 218
Single variable process/model/control, 147, 149–150, 152–154, 156–158, 161–162, 166, 170–173, 182, 191–192
Smith Predictor, 5, 99, 111, 171–172
Stability, 14, 17, 19, 24–27, 33–36, 39, 43, 45–46, 49, 51–52, 66, 84, 92–94, 101–102, 104, 109–111, 113, 120, 126, 132, 135, 139–141, 145–149, 152–156, 169, 174, 182, 190, 196, 201, 226, 231, 240, 257

State-space. *See* Models and modeling
Steam, 126, 134, 184, 192
Steel, 170, 172, 194, 195
Step response. *See* Process behavior
Supervisory control, 175, 240

Temperature control, 1, 15–16, 48, 77, 117, 126–133, 137–141, 175, 183–184, 191–195, 205, 225–226
Training, 5, 146, 191, 234, 253, 255–256, 263
Transfer functions. *See* Models and modeling
Tuning, 2–3, 5–9, 11–12, 15–17, 23, 25–28, 30–31, 33, 35, 37–39, 41, 43–46, 48–49, 51–53, 56, 64, 68, 70, 72, 86, 91, 99, 108, 110, 117, 127, 134–135, 139, 142, 147–148, 152, 157, 170, 174, 176, 182, 184, 186, 192, 196, 198, 203–206, 208–209, 211–212, 215, 217–219, 222–225, 229–231
 auto-tuning, 3
 detuning, 37, 48, 142, 148, 170, 173
 manual, 5, 145, 206
 on-demand, 204–205
 pretuning, 45–47, 49, 52
 rules, 15–16, 209, 218, 225
 self-tuning, 3, 45–48, 51, 56, 66, 70, 146, 189, 205–208, 210
 tests, 5, 7
 Ziegler-Nichols, 5, 206

Universal Dynamics Technologies, 3, 99, 142

Valves, 1, 15–16, 25–26, 47, 79, 126–127, 129–130, 135, 158, 180–181, 196, 219, 226, 229–230
Variance, 17, 31, 75, 134, 208, 212, 226

Widgets, 242, 246–247
Windup, 25, 114

Ziegler-Nichols. *See* Tuning